How to be an IT Grrrl

My Life in Tech

A memoir by

Carol Wapshere

Blue Tongue Press

Published in Australia by Bluetongue Press
Canberra, Australia

Website: itgrrrl.xyz

Title: *How to be an IT Grrrl – My Life in Tech*
Author: Carol Wapshere

Print ISBN: 9781764572903
eBook ISBN: 9781764572910
Memoir | Women | Science & Technology

NATIONAL LIBRARY OF AUSTRALIA

A catalogue record for this book is available from the National Library of Australia

Cover design: Alicia Grady, Struck by Violet. struckbyviolet.com
Book production: smartwomenpublish.com

This book is a work of memory, and my recollections may not match others'. Aliases have been used, dates are approximate, conversations have been reconstructed, and stories have been simplified for readability. However, everything that I've said happened to me did happen, and everything that I've said I did, I did.

While I hope that some people may find inspiration, validation or solace in reading about my experiences, this memoir is not intended to be career advice.

This book was written on Ngunnawal country.

Table of Contents

I'm an ITGrrrl:

BECAUSE tech work is brain work, and there's nothing intrinsically gender deterministic about having a brain that's suited to logic, creativity, and building stuff with computers.

BECAUSE a complex mess of assumptions and lazy thinking has led to the broad societal conclusion that Only Boys Really Understand Computers, and one way to fight this is to keep showing up day in, day out, still not a boy, and still understanding computers.

BECAUSE the gender pay gap is maintained by women being encouraged, cajoled, and shamed into 'lower value' jobs with less pay and more redundancies.

BECAUSE computers have infiltrated every aspect of our lives, not always in good ways, and the only way to limit the automated fibre-optic speed electronic fuckery is if these systems are commissioned, designed, built, and run by a representative cross-section of the community.

BECAUSE well-meaning Women In Tech programs often gloss over the technical, inadvertently reinforcing the notion that women may be *around* the tech, but they don't *do* the tech.

BECAUSE too many people in tech still believe bullshit over track record, and it's rarely the women peddling the bullshit.

BECAUSE everyone should have an opportunity to do the type of work that suits their unique self, for fair and equitable compensation.

BECAUSE it is important to see diverse people doing all types of work, and for us to support each other in doing so.

With thanks to Kathleen Hanna and her RIOT Grrrl Manifesto.

1. How to be an IT Grrrl

IN JULY 1991, IN THE CITY OF OLYMPIA IN WASHington State, USA, fifteen women gathered. The meeting had been called by Kathleen Hanna from the band Bikini Kill—whose manifesto I repurposed into my own IT-flavoured one as the introduction to this book—and Allison Wolfe from the band Bratmobile. Their goal was to create a space for women to discuss their experiences of sexism in the punk scene, but it would lead to something much bigger.

This meeting was the inception of Riot Grrrl: a leaderless, do-it-yourself movement that inspired young women around the world to channel their anger, individuality, and personal experiences into rebellious creativity. Riot Grrrls started bands, produced zines, and made art. The ethos was avowedly feminist, inherently political, and defiantly unruly.

At that time, I was embarking on adulthood and figuring out who the heck I was after a rough and confusing childhood. When I first heard about Riot Grrrl, the visual image it conjured of girls kicking down doors that should never have been closed appealed to me greatly. It still does today.

In 1991, I couldn't have told you where Olympia was, but I was well aware of another city in Washington State. Seattle seemed to be where everything was happening, both in the grunge music that had so strongly gripped my twenty-one-year-old ears, and in the personal computer industry that was shaping my nascent career in Information Technology.

I wonder, had I teleported myself from my hometown of Sydney, Australia, into one of those Olympia meetings, what the original Riot Grrrls would have made of me. I looked the part, with my unnaturally coloured hair, my dangly mismatched earrings, my charity shop clothing, and my Blundstone boots. I loved the music, especially when it was loud and live—I'd even been lucky enough to see Nirvana when they toured Australia in early 1991. I played in an unknown indie band myself, though in what will be a repeating pattern in this memoir, I had found my own little niche within that scene. When I got up on the filthy and liberally gaffer-taped stages of the inner-Sydney band pubs, I was the only person I knew doing it with a violin in their hand.

In the music side of my life, I shared some similarities with those original Riot Grrrls, but in my work life, I was very different. I might have occasionally played at being an indie musician, but my real vocation was Information Technology. Instead of hustling for gigs or creating art, my days were spent managing servers, running network cables, setting up PCs, installing applications, and fixing computer faults. And I loved it—so much so that now, in 2025, I still work in IT.

It is with a strange mix of pride and disbelief that I reflect on how long I've been in the tech business. I started all the way back in 1985 with a summer job for Apple Computers when I was only sixteen. In the forty years since, I have worked in four different countries, for multiple employers, on countless IT projects, and with an untold variety of hardware and software. My IT career has supported me through international moves, raising children, and the usual ups and downs of life. It has been, on balance, a very good career.

Statistically, my gender makes me a rarity in this business. Of all the white-collar professions, computing has had, during my time in it, by far the worst track record for attracting and retaining women. The women who are in IT are many times more likely to do non-technical work, making me even more unusual for staying on the technical track. I have bucked the trends—but for a very long time, I didn't give much thought to why that might be.

It wasn't until those strange years of the Covid pandemic that, in dire need of distraction, I looked back over my career in detail, searching for clues. Why was I so ready to dismiss the incidents of bias I copped as one woman among a hundred men? What gave me such determination to push through? How, when things went bad, did I get back up, dust myself off, and keep going?

I spent so much time teasing out the reasons that I wrote this memoir. And in doing so, I spotted a thread that ran through it all—one that shares its philosophic DNA with Riot Grrrl. It is my sense that conformity is a trap, my belief that I have as much right as anyone to speak, and my preparation to argue my position, no matter how 'unlikeable' that makes me.

Like the Riot Grrrls, I didn't follow the scripts I was given; I wrote my own. For me, this meant building a career in the rapidly developing and intractably masculine world of IT, where there were no paths to follow but my own. I have been, I decided, with satisfaction, an IT Grrrl.

*

In 2023, as the world around me emerged from the trauma of the Covid years, something happened that brought my IT Grrrl concept together in an unexpectedly perfect way. It all started one day when I spotted a social media post from my favourite local music venue about a 'Public Service Battle of the Bands'. Run by the (completely fictional and not at all government-affiliated but awesomely named) 'Department of Rock', it was a competition for adult, amateur bands.

I should mention that I now live in Canberra, which, in case you weren't aware, is the capital city of Australia. Canberra is very much a public-service town, with the federal and territory governments employing a substantial percentage of the population. While not a public servant myself, I am one of the legion of IT contractors and consultants who work across the Government Departments and Agencies. And according to the rules of the band competition, that made me eligible to enter.

Which suddenly was the one thing I wanted to do more than anything else. On reading the post, an idea had popped into my head and grabbed me hard. 'How cool,' I thought, 'would it be to enter a band made entirely of women who work in IT!'

I didn't know if it would even be possible. Back in my band scene days of the early nineties, it would not have been—I had known exactly one woman who worked in IT, and that was me. But the sector had exploded since then and there were many more women now...

I obsessed over the idea for a few days, then I posted on Facebook:

```
Bit of a weird request here. I want to put
a band together for the Department of Rock
band competition and I am specifically
looking for people who:

   - Identify as female,
   - Work in IT in the public sector,
   - Can play an instrument or sing, and
   - Think this band comp sounds like fun.

If that's you or someone you know,
please DM me!
```

OK, I thought after clicking POST, my idea is out there in the world. If no one replies, then I'll know it wasn't meant to be.

By the next morning, I had a bass player and a guitarist. My crazy idea had legs!

I invited both women over so we could meet each other and make plans. Sharon, the bass player, was the IT Finance Director for a major Government Department. Serin, the guitarist, worked for a tech vendor, embedded in another Department as a Services Delivery Manager. They both loved the concept of an all-women-in-IT group.

'I might know a drummer,' I told them, 'but we're going to need two singers. I have this idea to turn "Wannabe" by the Spice Girls into a sort of argument between a person on the service desk and a demanding caller. You know, telling them what they really, really want, and then it's like a new laptop or something?'

Both women laughed, which was very encouraging, as at that point I had only tested my ideas on my husband Nathaniel, and I didn't know if anyone else would get my particular sense of humour.

Within a couple of days, we had also recruited Kiara, a young developer with an angelic voice, and Monica, a website designer and drummer. That just left one more singer to find.

I was asking around at a public-sector office where I regularly work when someone mentioned Charmaine. 'I heard her sing at a wedding a few years ago; she has an amazing voice,' my source told me.

I knew Charmaine. We were both technical specialists, contracting through the same small IT services company where we were the only women among thirty-odd consultants. I knew a little about her work history as a data scientist, but I had no idea she'd once been a professional singer. Which goes to show you can't make assumptions about people just because they happen to work in tech. Something I should know!

I messaged Charmaine about the project, asking if she could join us on vocals. 'Let me think about it,' she replied.

Meanwhile, the rest of the band got together for a read-through of my rewritten lyrics to 'Wanna new laptop'. I also presented 'They tried to make me go to Training' (a la Amy Winehouse), and 'Pretend we can Tech' (from a song by nineties grunge legends L7). My new bandmates were gratifyingly enthusiastic, but agreed we needed a second singer.

I pinged Charmaine again, this time sending her my rewritten Rehab lyrics. She replied five minutes later with a video message of her singing my version in a voice that could have given Winehouse a run for her money—I had to have Charmaine in my band!

However, she was still reluctant to commit, and I got it. She had two young kids and, in her already limited spare time, hosted a podcast about women in data technology. She reminded me of myself fifteen years earlier, when my kids were young but I'd felt an intense need to push myself—determined to be a good and present mother while also successful in both my day job and the unpaid IT community work I had assigned myself. It didn't leave much room for hobbies.

But I couldn't get the sound of Charmaine singing 'Training' out of my head. I rang her and made my final pitch.

'Look, I know you're stretched, but this competition is going to be amazing. It's such a cool way to represent women in tech—I think you'd really enjoy it.'

'Oh... um... I don't know... but it does sound... Oh damn it, I can't let my IT ladies down,' said Charmaine. 'Right! I'm in! I'll make it work!'

And that was it, I had a band! And of course there was only one possible name for the group. Together, we became the *IT Grrrls*.

As we rehearsed over the following weeks, I looked around in delighted disbelief at these women I had assembled. Just by chance, we represented a neat cross-section of women in IT. We had different types of jobs: Sharon and Serin on the management side, Kiara programming, Charmaine in data, Monica in media, and me in backend systems. Our ages ranged from twenties to fifties. Some of us were mothers, at different stages of our parenting journeys. All of us had a variety of interests which somehow, by miraculous good luck, included playing in this band.

Most importantly, we gelled immediately, with friendship soon following. I felt immensely grateful to these women for so whole-heartedly embracing my ideas. I couldn't wait to get up on stage with them and prove our mission statement that *Women who work in IT are total Rock Stars!*

And if you want to know how we went in the competition, you're going to have to stick with me until the end of this book. Or you could sneak a peek now, it's up to you, but please do come back. I'll be waiting right here, and I have lots more stories to tell.

2. How to learn from history

I WORK IN AN INDUSTRY THAT IS NOTORIOUSLY bad at learning from its own short history. Projects fail for exactly the same reasons as thousands before them. Workshops are held to identify the 'Lessons Learned' from the last failure, and then nothing changes. Attention and resources are diverted to the Next Big Thing, while the Last Big Thing and the Big Thing Before That are left unfinished.

I don't mean to imply that everyone in tech is incapable of learning from history, in fact most of my technical colleagues are very good at it. We try hard to replicate what worked before and avoid what didn't, but we don't always get to decide. We are often at the mercy of decision makers whose relentless focus on the future is mirrored by their reluctance to look behind. Uber's intention to 'move fast and break things' summed this up neatly. That this was widely considered an admirable business plan exposes a wilful blind spot when it comes to the value of the past.

On the history of women in computing, this blind spot is so big it looks like erasure. I have a formal education in Computer Science, and I have spent my entire adult life immersed in the tech world, yet it was only in recent years that I learnt about computing's female history. And when I did, I was astounded to read that computer programming was once a *female-dominated* profession.

The very first modern computers, built in the USA in the 1940s, were programmed almost exclusively by women who had to invent their own methods to do so. In the 1950s, as the mainframe business grew, programming was broadly considered 'women's work', akin to typing or operating a telephone switchboard. In the 1960s, the most successful software firm in the UK was almost entirely staffed by women writing code from home. And female computer scientists have developed powerful algorithms, whole programming languages, significant methodologies, and innovative approaches that have advanced computing across the range of technology.

For a very long time, I knew nothing of this history. In my 1980s adolescence, movies and TV had shown me a male computer nerd stereotype, and this was reinforced when I started university and saw only one woman on the Computer Science faculty. It was easy to assume that computing had a masculine past, but I implicitly believed that it had a gender-balanced future. This, I thought, was both inevitable and close. In my short life, I had already seen such progress towards equality that it felt unstoppable.

I remembered how, in the 1970s, I'd heard both women and men saying they could never trust a female doctor. And at election time, when I tagged along to the voting booth with my mother, all the faces on the placards were male. I remembered listening to talkback radio sometime in the late 1970s as people of both genders phoned in to protest the recent addition of female DJs to the station's roster. 'I'm not sexist,' these callers said, perhaps even believing it, 'but a woman's voice just doesn't sound right on the radio.'

When I started my degree in 1987, I believed that society had moved on a lot since those unenlightened days of the 1970s. We now had female politicians and radio announcers, and female GPs often had the longest waiting lists. Also, in possibly the biggest hurdle for equality in Australian culture, there were now women reading the sports results on the evening news! Surely the number of women in computing was also going up.

Spoiler alert, it wasn't. Quite the opposite, it was going the other way.

I had no way of knowing that female enrolments in Computer Science had peaked in the mid-1980s, a year or two before I enrolled. The long slide down, from which we are yet to fully recover, had begun. My first-year class in 1987 was around thirty percent girls, but had I enrolled in Computer Science a decade later, I would have been lucky if there had been half that number.

And it wasn't just education: as the pipeline of female graduates shrank, and the industry became increasingly 'tech bro', the percentage of women working in computing followed the same downward trajectory. In 1989 I had a job where I was one of three women in a team of ten programmers. Ten years later, I would have been the only one.

In 1991 I got my first IT systems administration job. I like to claim that I didn't know how male-dominated IT was back then because I was half of a team of two with a lovely chap named Geoff, so from where I sat, IT had perfect gender parity. But I'm fudging a little: I knew computing jobs were male-dominated, I just didn't expect them to stay that way for so long.

By the second half of the 1990s, I had to admit that the young women I'd expected to swarm into IT behind me were not showing up. At this point, I might have been curious about why that was, but instead I actively avoided the subject. You might have too, had you been repeatedly asked to answer for your entire gender.

'Why are there so few women in IT?' someone (or to be more precise, some *man*) would ask me, completely out of the blue.

Here we go again, I'd think. I might shrug non-committedly in the hopes they'd leave me alone, but if I was feeling argumentative, I'd snap, 'What are you asking *me* for? Go talk to the women who *aren't* here.'

It didn't matter what I said, it never stopped the pet theory that was coming next. 'Girls think computers are boring,' I might be told. Or, 'Women are put off IT because they think we're all nerds.' Or the classic, 'Men are naturally better with computers because we like tinkering with machines.'

I wish I'd known then to say, 'Listen here matey-boy, women have been programming computers since there were computers—what have you got to say about that?'

But instead I'd just mumble, 'Well I'm a woman and I'm here...'

'Oh you, you're different,' they'd say.

Gee, thanks.

I mean, OK, I knew I had my differences, but I didn't think an ability to build a computer system while in possession of lady parts should have been one of them. I reasoned that if I could work with computers, there was no reason why loads of other women couldn't as well.

I was, I see now, discounting my own history.

*

My colleagues were right, I was different, and I knew it. I had been brutally apprised of this knowledge in primary school grades three to six, when I was seven to eleven years old. I was *that kid*. The one who was always alone. The one who my classmates felt free to publicly despise, humiliate and, on occasion, physically assault. The one who no one, no child, teacher or even my own parents, ever protected.

For those years my name wasn't Carol, it was 'Smellbomb'. This renaming originated with a particularly power-hungry and well-connected girl in my class—she was the granddaughter of the Principal, and in the social hierarchy of the school, as good as royalty. What started as a taunt from her inner circle progressively spread throughout the class as the example of my ostracism made clear what would happen to anyone who didn't fall in line. Our princess had ordained me a despicable and untouchable Smellbomb, and no one, not even a series of class teachers over four years, challenged her on it.

The humiliations were daily, starting first thing each morning with the class line-up. We were supposed to form two equal lines, but instead there would be one with twenty-nine kids and one with just me. A teacher would chastise 'Don't be silly' until some kid loudly announced their bravery and came to stand next to me, protecting the others that followed from my contamination. In class I sat alone in the back corner because no one would sit next to me. At recess and lunch, I went to the library or joined the few girls from another class who tolerated my presence. I'd sit with them and watch them eat their lovingly packed snacks and lunches while I went hungry, for reasons I'll get to later. And then, at the end of each break, I had to again join my vile classmates for a reprise of the line-up 'silliness', followed by yet more ostracism and abuse.

It was psychological torture of the type I wouldn't wish on my worst enemy, and I copped it every school day from age seven to eleven.

Inevitably, those years inflicted substantial damage on my sense of self and security—so it pains me to admit that there was one career benefit to come out of it: I am very used to being the odd one out. In my twenties I even became proud of it, wearing my differences like a badge of honour, my status as a rare 'woman in IT' the gilded crest on top. For my inner Smellbomb, that bullied little girl who still lives inside me, my unusual career in IT has been a finger up to the shadows of the jeering mob, proving that I, unlike them, can think for myself.

Compared to the daily humiliations those little shits from primary school put me through, being a woman in IT has been a breeze. If people looked surprised when I introduced myself as the technician, I shrugged it off. If they felt the need to question me closely on my credentials, I took it in my stride. When I had to work harder than the men to prove myself, I refused to let it bother me. These were but trifling annoyances, I told myself, I'd survived far worse.

Thankfully, the life lessons I learnt in high school were far more positive—I honestly don't know where I'd be today if I hadn't gone to that school. It was a well-regarded High School for Brainy Girls, and difficult to get into—but while the primary school bullies may have made my life a misery, they couldn't stop me blitzing the selective schools entrance exam. Being offered a place in that high school was all the sweeter because I knew none of those brats would be there.

At high school, I found friendship among bookish nerds like me. Maths and science were actively promoted, with teachers often reminding us that *Girls Can Do Anything*—a message I took deep into my heart.

A huge bonus of my single-sex school was the lack of competition with boys during those awkward adolescent years. When a couple of friends and I wanted to spend our lunchtimes playing with the school's brand-new Commodore 64 computers, we didn't have to fight our way through a gaggle of sharp-elbowed boys to get to them. Thanks to that, my first experiences with computers were easy, fun, and criticism free.

*

I thought a lot about history during the Covid years. In 2021, for reasons which I'll get to later, my mental health deteriorated in a way it never had before. I was in dire need, experiencing intense flashbacks, my brain in turmoil, my body wracked with stabbing pains—but I was far from the only one suffering during the pandemic, and it was impossible to find an available psychologist.

At the time I was working on two projects for two new customer organisations, while also supporting another of my long-term ones. I had my family to think about—my husband, two teenagers, and my mother-in-law, who was living with us and terrified by what they were saying on the news about this virus and its impact on the elderly. Not wanting to burden my family I confined my breakdown, as much as possible, to the privacy of the backyard studio I was using as my work-from-home office.

But the hurts and the painful memories did not agree to this confinement. They grew more insistent until every conversation I had, every bit of technical analysis or solution development I did, every board game I played or movie I watched with my family, had to share brain space with screaming anguish.

Weeks of this torment turned into months, but I don't give in easily. Without access to my usual coping mechanisms, live music in particular, I looked for ways to help myself at home. Out of sheer self-defence, I started to write.

Later, when we get to know each other better, I'll tell you some more about the nasty memories I was disgorging onto the page at that time. Writing them out was no quick fix, but it did help. Over time the turmoil in my brain calmed enough for me to turn my thoughts to things more uplifting. I continued to write, but it became about my family, my dearest friends, my love of music, and my career.

As I jotted down the anecdotes of my working life, I thought about the question that had vexed me for so long: why were there so few women in computing? Covid had cancelled all my other plans and I needed a distraction—perhaps now was the time to look for answers.

So, I read: books, Wikipedia pages, articles, Reddit threads, and research papers. For the first time I learnt about Grace Hopper and the ENIAC Six and Margaret Hamilton and Kathleen Booth, and many other women who literally invented the art and science of programming. But this just deepened the mystery: large numbers of women had, back then, done technical work, and very capably. So, what happened?

The first clue came from Nathan Ensmenger's book *The Computer Boys Take Over*. He describes how, in the 1960s, the burgeoning computer industry was so short of qualified people to program its expensive and hulking mainframes that it was dubbed the Software Crisis—NATO even convened a special summit in 1968 to discuss it. By then, the skills shortage had increased the pay and status of programming to the point where it flipped from 'women's work' to 'men's work'—a gendered recharacterisation reinforced by widely used aptitude and personality tests. Thanks to some very specific biases in these tests, they prioritised 'antisocial, mathematically-inclined males'. Profiling, as Ensmenger says, 'created a gender-biased feedback cycle that ultimately selected for programmers with stereotypically masculine characteristics'.

It was a shock to realise that the male computer nerd stereotype I had unconsciously absorbed was never a naturally occurring thing. It was a fabrication, created by biased profiling, and then perpetuated as the selected men designed the personal computers and games that most appealed to their younger selves. It is hardly surprising that the types of people this tech was designed for were also the types of people most attracted to it.

I was very lucky, I realised, to have studied Computer Science when I did. The male computer nerd stereotype was a thing in 1987 when I started university, but widespread home computer ownership was not, and because of this, there was no assumed knowledge in my first-year classes.

As I read in Jane Margolis's *Unlocking the Clubhouse*, a book based on her research into the Computer Science program at Carnegie-Mellon University in the late 1990s, things changed a lot in the decade after I did my degree. Faculty members expected first years to arrive on day one with advanced skills, and some were openly dismissive of students not prepared to live and breathe computers. Their attitude was particularly off-putting for young women who had grown up in a time when the male ownership of computers was established in society. As Margolis says, women were more likely to 'question their ability to become computer scientists because their gender has ... rendered them suspect'. I started to understand why, in the 1990s and 2000s, I so rarely saw younger women entering my profession.

It had been very enlightening to read about the experiences of the women who came before and after me in computing, but it was *The Athena Factor* that hit me right in my lived experience. This book came out of an early-2000s study commissioned by a consortium of household-name companies employing large numbers of scientists and technologists. They wanted to understand why the quit rate was so high among their female employees, with the computing professions having, by far, the worst track record. I had long maintained that I had rarely, if ever, encountered bias in IT—and yet I immediately recognised almost all the discriminatory situations described in this book. So often had I encountered clearly biased treatment, and yet I had kept brushing it off, telling myself that it didn't matter.

This is when I finally faced something I had long suspected but had shied away from acknowledging: I have survived in this industry as long as I have because the primary school bullies trained me for it.

From a young age, I had no choice but to tolerate scorn and abuse. To survive, I had to brush it off and keep showing up, day after day, all the while knowing it would be hell. And in doing so, I found something that muted the abuse: learning. If a topic really interested me, if I could focus intently on a classroom task, the bullies couldn't touch me.

I can no longer deny that this has helped my career. It allowed me to ignore all but the most egregious slights, it drove me to continue showing up in difficult, even hostile situations, and it primed me to find purpose and satisfaction in challenging and absorbing work. But I have to say, if a woman must be as messed up as me to survive in tech, then the problem is clearly the workplaces and not the women.

We are all products of our history, including those of us who work in an industry as future obsessed as technology. I'll have a bit more to say about my early history and how it has affected my career, but first, something more directly pertinent. I'm going to take you back to my earliest jobs in the computer industry.

3. How to experience work

BEFORE I READ UP ON THE COMPLEX FACTORS deterring young women from IT careers, I thought the problem was a simple lack of information. Sometimes I told teenaged girls and young women about the wide range of interesting jobs available in IT, but they never seemed interested. I wished they'd just have a go, get some work experience, find out for themselves. I was sure they'd find something that suited them because this was, quite by chance, what I had done.

The promotions one

I got my first job in the computer industry in 1985 when I was sixteen years old. It came about after someone rang my school looking for a teenager who liked computers but wasn't overly shy. Their call was routed to the teacher who had run the school's first ever programming class, one that

I had taken. I guess my combination of quick uptake and loud mouth had stuck with him and he came looking for me.

'Someone rang the school yesterday, from Apple Computers, have you heard of them?' he said.

I shrugged; it sounded familiar.

'Anyway, they're looking for teenagers for some kind of summer job, and I thought you might be interested.'

I was very surprised and, I won't lie, quite flattered. The teacher handed me a piece of paper with a name and a phone number on it, and I couldn't wait to get home and make the call.

'We're looking for teens to demonstrate Apple computers at department stores in the lead up to Christmas,' explained the woman who answered the phone. 'It will be Wednesday to Saturday for four weeks, with some training beforehand. Are you interested?'

'Yes I am!' I replied. I didn't have the slightest clue what the job might entail, but I didn't care. She was offering me money and an excellent excuse to stay out of the house during the school holidays—double bonus!

That call was my only interview, and apparently, in expressing my interest, I had answered the only question correctly. The woman signed me up on the spot.

A few months later, on a sunny November morning, I arrived at the corporate office of Apple Computers. This was long before glitzy Genius Bars and shiny glass headquarters, and Apple's Sydney outpost was a grey, nondescript building in a semi-industrial area, overshadowed by the concrete pylons of a flyover road.

I was directed to a training room where I met a dozen or so other teens, perhaps a couple more boys than girls.

We milled nervously about until two adults appeared and introduced themselves as our trainers.

'Before we get started,' they said, dimming the lights, 'we'd like you to watch this.'

We found seats at the empty desks in the room and for the next five minutes we watched an infomercial about the marvellousness of Apple the company and the total and utter brilliance of their Series II home computer. After the lights came back on, it took me a moment to steady myself. It wasn't just that my eyes needed to adjust to the sudden brightness of the room, but that I myself needed to adjust to the sudden brightness of a world which contained The Apple Computer (cue choirs of angels).

We were each given a boxed Apple IIc to unpack and assemble. It felt like Christmas as we prised open the boxes and gingerly removed the packaging.

The IIc was Apple's first attempt at a portable computer, with a casing that was deliberately reminiscent of a briefcase, and a handle that doubled as a stand. The keyboard was built-in, but the monitor was a separate and weighty accessory—hardly something you'd tuck under your arm if you were on the go. Instead, our trainers explained, Apple's mobility solution was a cable that could be plugged into a television set. At that time, I lived in a one-television household, so while I accepted the idea that this cable made the IIc theoretically portable, I did wonder what you were supposed to do if your dad wanted to watch the tennis.

Our mission, over the two days of training, was to learn enough about these computers to be able to demonstrate them, and ideally sell them, in Grace Brothers department stores. Apparently, our fresh-faced youth would convince the buying public that these daunting and expensive machines were as easy as child's play.

But first, we children had to understand a few things. Other than some very basic BASIC skills, I knew almost nothing about computers.

'What's RAM?' I asked, after hearing the word several times.

'What, don't you *know*?' sneered one of the boys.

'Now now,' chastised the trainer, 'there are no bad questions.' He then gave me an explanation of RAM that left me almost exactly as clueless as before.

Training flew by, and already it was time to report to my assigned department store. I was still sketchy on many details, but if there was one thing I was absolutely certain about, it was that Apple was The Greatest Company Ever!

The computer section manager in the Bondi Grace Brothers took one look at me and my Apple name tag, Apple pendant, and Apple earrings, and said 'I see you've been through the brainwashing.' I was confused, didn't he know that Apple was the best?!

My job was to stand beside an Apple IIc and show off the future of portable computing to anyone who might stray within hailing distance. There were long idle periods waiting for punters, and rush times when I'd explain to one disgruntled customer after another that I couldn't help with any non-Apple enquiries. Every now and then, someone agreed to hear my spiel about the IIc.

Invariably, they asked questions I couldn't answer. 'How will this help my child with their schoolwork?' was a common one, or 'Why should I buy this instead of the

Commodore 64?' I found the latter particularly troublesome—it had a similar spec to the IIc but, at $640 (AUD), was a third of the IIc's $1800 price tag.

I phoned my Apple trainer to relay questions and ask how I should answer next time. 'Tell them it's a better-quality product,' he said, 'and with much more software to choose from.' But this was not supported by what I saw on the shelves: there was very little software for the Apple II, and it was much pricier than Commodore equivalents.

After four weeks, the demonstrators were recalled to the office for a wrap-up. We hadn't seen each other since the training and asked, 'How was it? Did you sell any?' I had not sold a single computer but barely anyone had. The only kid with any success had been in the flagship city store. He told us that a couple of Apple IIc computers had sold, but that was nothing compared to the $8000 Macintoshes that were walking out the door. 'I didn't have to do anything,' he admitted. 'People just wanted to buy Apple computers, and they didn't care about the price.'

Despite our lack of sales, the trainers enthused about the project's success, assuring us we'd all done very well.

'We have a special surprise,' one of them said, grinning broadly. 'We are now going to award an Apple IIc to the best demonstrator!'

Two cardboard boxes containing a IIc and a monitor were brought into the room. We all glanced towards the city-store kid—obviously, he was the success of the group and would get the reward.

I can't have been the only one who was shocked when they called out the name of the girl who hadn't even known what RAM was.

It took a second to register that they were really giving me the computer. Dazed, and avoiding eye contact with City Store and the other politely clapping kids, I floated to the front of the room. I wanted to hug the boxes, but limited myself to stroking them, confirming they were really real. A taxi was summoned so I could get the boxes home, and I was still in a state of excited disbelief when I dragged them in through the front door an hour later.

My father helped me unpack the computer, and we set it up in my bedroom. He left me to it as I inserted the floppy disk of the only game that came with it: a text-based adventure based on *The Hitchhiker's Guide to the Galaxy*, a book I had belly-laughed my way through many times. I was in the early stage of the game—having, as Arthur Dent, just left the bulldozer and gone to the pub with Ford Prefect—when it was dinner time in the real world, and I had to leave my new toy. When I returned, the computer was unresponsive. I rebooted it and started the game again, but it froze almost immediately. 'We've both had a big day,' I surmised, and decided to try again tomorrow.

Over the following weeks, the computer continued to freeze at random intervals. Sometimes it happened almost immediately, and at others it ran for a while but then got stuck just as I was trying to figure out what a 'buffered analgesic' was, and why the Hitchhiker's game wanted me to do something with it.

The warranty card said that if there were problems, you should take the computer back to the place you'd purchased it, but I hadn't purchased it. I asked my father for help, but he brushed me off—he later admitted that he'd assumed, as we sometimes say a little meanly in IT, that the problem was in the chair and not on the desk. By the time I managed to show him what was wrong with the computer, the miserly three-month warranty period had expired. I was stuck with an expensive paperweight, and I never got to finish the Hitchhiker's game.

By then, I was completely over any vestiges of Apple brainwashing. I was left with an impression of over-priced, over-marketed products that didn't deliver on their promise of quality. Say what you will about how Apple revolutionised the user interface, popularised the smartphone, and elevated personal computing to must-have consumer products through clever and elegant design ... I just can't buy it.

The experience taught me much more than an abiding suspicion of Apple. For instance, I learnt that I don't have a flexible enough relationship with the truth to be any good at promotions and sales. When asked a direct question like 'How will this help my kid?', I could only answer honestly that I didn't know. I couldn't even bring myself to repeat my trainer's assertion that there was more Apple software to choose from when the evidence on the shelves directly contradicted it. Perhaps I may have sold a computer or two if I'd been prepared to tell a story that was convincing rather than strictly honest, but I knew now that just wasn't me.

And yet, despite my lack of sales, the Apple people had seen something in me. I guessed that I stood out because I kept asking questions—I showed that I wanted to learn.

The computer they gave me might have been broken, but being deemed worthy of it was the real prize. A year later, selecting Computer Science as one of my university subjects felt like a no-brainer.

The teaching one

My second computing job came right when I most desperately needed it. I was eighteen years old, had recently started second year university, and was facing a stark new reality.

A few weeks earlier, I had been living with my father, his wife, and their two children. My relationship with my stepmother had never been good—not since two days after the wedding when she, very clearly and deliberately, told thirteen-year-old me that she intended to get me out of my father's house by any means necessary. We'd had some rocky years since then, but after my baby sister and brother came along, my usefulness as a live-in babysitter bought me some forbearance. Until, that is, early in second year university. Without warning, she kicked me out, screaming, 'And don't ask your father for any money!!', as she chased me out the door.

This wasn't the first time she'd kicked me out on some pretence or other. As usual when this happened, I ran round the corner to my old school friend Emma, and her family kindly gave me shelter from the stepmonster storm. On past occasions I had eventually been allowed to return home, but this time felt different. I was eighteen, officially an adult, and I'd had it to the back teeth with her bullying. It was, I recognised, high time I moved on.

I found a room in a student share house, and I drew up a budget which made one thing clear: even if I stretched my savings as thinly as possible, I'd barely make it halfway through the year. I needed a job.

A couple of days later I was in the computing lab unburdening my troubles onto a friend when he surprised me with a helpful suggestion.

'I heard the School is looking for second year students to work in the first-year programming labs,' he said. 'A couple of guys I know are doing it. You should go see if they need anyone else.'

He told me the name of the staff member in charge, and I went straight to his office.

This man was a well-known curmudgeon, as I knew because he was my Digital Logic teacher. I enjoyed the lab work in his class, which involved wiring up giant circuit boards, but I could have done without his cutting criticisms and cynical asides.

The Curmudgeon, as I will call him, looked more annoyed than usual at my sudden appearance in his office, but he at least heard me out.

'OK,' he sighed with clear reluctance. 'There's an information session for the new lab supervisors at lunchtime today. You can come along, but I'm not promising anything.'

At the meeting, I found I was the only girl among eight or nine boys. The Curmudgeon wasted no time on niceties and got straight down to business.

'All shifts are one hour long, I will allocate them and there will be NO COMPLAINING,' he said. He didn't look in my direction, but I felt this comment was directed at me.

He asked for a show of hands for each shift as he read them out. I raised mine for every shift, and every time the Curmudgeon gave it to one of the boys. Some of them were even getting second shifts while I still had none.

Then something changed. The Curmudgeon read 'Monday 5 to 6 pm', and I was the only one to put up my hand. And then he read 'Tuesday 8 to 9 am', and again, I was the only taker. And on it went: any shifts at either end of the day were of no interest to my comfortably supported classmates, while I'd take anything I could get. By the end of the session, I had more shifts than anyone.

My entire training for this job took no longer than three minutes. On my first day, I was rostered on with a third year who led me to a free computer and showed me how to open the bottom of the mouse, remove the ball, and scrape out, with my fingernail, any grime or dust that was blocking the tracking wheels. It was my first troubleshooting technique, and I needed it as those desks were filthy and the mouse wheels were always getting clogged. 'Excuse me, my mouse isn't working,' was the most common request I got.

Otherwise, a first year might ask for help with their code. They were working through the same book that I had the year before, doing exercises I could well remember nutting through myself. When a student asked me why their code wasn't working, I never directly pointed out their mistake, even when it was so obvious it might have had flashing neon signs around it saying BUG HERE! I knew it was important they understand the problem, so I guided rather than told. I loved it when they said, 'Ahh! Got it!'

Over time, as he noticed how dependable I was, the Curmudgeon softened towards me. I always turned up for my shifts on time, and I never cancelled on short notice like some of the others—and when they cancelled, I was ready to fill in. I made his life easier, and he liked that about me. In fact, it got to the point where he seemed to think we were friends, and he'd share conspiratorially snide comments about colleagues or other students. He was a creep and a bully as well as a curmudgeon, but I needed the job, so I kept those views to myself. I maintained a blandly friendly exterior and continued to accept any shifts he sent my way.

Perhaps he did me a favour. In the years since, I have occasionally come across a co-worker who reminded me of the Curmudgeon. They were suspicious and disparaging, their minds already made up before I'd even opened my mouth, but I always believed they'd soften once they saw what I could do. And in almost every case, that's exactly what happened. (Only one seemed entirely incapable of ever accepting me, and that situation turned very bad—but we'll get to him later.)

At the start of my third year of university, I received a surprising payoff for my diligence when the Curmudgeon said he'd recommended me for a casual position teaching first year tutorials. This was not something that was generally offered to an undergraduate, but once again there was a lack of takers for the early and late slots. I was happy to have them. Tutorials paid double the hourly rate of labs, and I got extra for preparation and marking. I still had to count every cent, but at least I now had more of them to count.

I felt very fortunate to have landed those teaching jobs—not only were they conveniently on-campus but they paid several times the hourly rate of the retail and hospitality jobs that other students did.

But while teaching was something I did reasonably well, I knew it was not for me in the longer term. I didn't want to teach other people how to write programs, I wanted to get out there and write them myself. While still an undergraduate, I had an opportunity to do just that, and it showed me how satisfying real-world coding can be.

The factory one

In the second half of 1988, during my second year of university, I turned my thoughts to the long summer break ahead. My lab tutoring work would dry up before exams, and I needed a way to support myself until the next academic year.

For reasons I cannot recall, I became fixated on the idea of getting a summer job in a mine. Was it a yearning for rugged landscapes and endless horizons? Was I thrilled by the idea of the giant machinery? Or did I have some notion of testing and extending myself? Whatever the reason, it was enough to send me to the place you went for information in 1988: the library.

The university library had copies of the 'Yellow Pages', the business phone directory, from all Australian states and territories. I searched under 'M' for mining and minerals and wrote down names and addresses for companies across the country. I drafted a letter 'To Whom It May Concern', introducing myself as a second year Science student and asking for a summer job, anything at all. I photocopied it twenty times and folded each copy into an envelope. Stamps affixed, off they went in the post. I had no idea if I'd hear back from any of them.

I only needed one reply, and I got it—not from a mine in the distant Pilbara but from an aluminium processing plant on the outskirts of Sydney. They had a little programming job that a Computer Science student should have no trouble with, and they were offering me a short-term contract to do it. I hadn't imagined working somewhere I could catch a daily train to, but I wasn't going to turn down an offer.

I rang straight away and agreed to a ten-week contract, running from late November 1988 to February 1989.

I'm going to take you back there now, to nineteen-year-old me going to my job at the aluminium factory.

WEDNESDAY 18TH JANUARY 1989, SYDNEY

I stare out the train window as the suburbs rush by. I can't concentrate on my book because I'm thinking about the day ahead. Do I have all the logic right? Will the charts load properly? I'm worried I might have got the colour-coding wrong in one of them—better check that first thing.

The train arrives at Yennora, and I get off with a handful of other people. They will all be going the same way as me because there's only one reason anyone gets off at this tiny station. We follow each other to the end of the platform and down a ramp where we wait for the train to pass before crossing the tracks and entering through the gates of the Comalco Aluminium plant.

A couple of people go into the corporate building, just inside the gate. I went there on my first day to fill in paperwork and get told about safety and stuff, but I haven't been back since. I keep on walking down the road, passing various other buildings. I don't know what goes on inside any of them, but as I pass, I hear the sounds of machinery, warning bells, and the occasional shout. I keep walking towards my destination at the far end of the site—North Remelt. It's a massive warehouse, the size of an aircraft hangar, and it's filled with metal and machinery and noise and heat.

I'm not allowed inside without the right safety gear, so the first thing I do is go to my peg, just inside the entrance, where I change into my assigned shoes. They're very ugly, clunkier than the Clarke's school shoes I wore in the '70s, and a horrible baby-poo-brown colour. I'd love a pair of sturdy work boots like all the men wear, but I'd have to buy them myself and it's not worth it just for the couple of months I'm here. Anyway, as long as my shoes are steel capped, no one cares what they look like.

I grab my hard hat and stick it on my head. Then I follow the yellow marked pathway into the factory.

The path takes me past one of the crucibles. These things are amazing! They're massive, they must be two storeys high and filled with molten metal. It's literally like the inside of a volcano in there! I walk by at a safe distance, but I can feel the heat radiating from it. It looks incredible when they tilt it and pour out the molten metal. I can't be on the factory floor when that happens, but I've watched it from the big window in the office.

I reach a flight of galvanised steel stairs, and I climb up them to the office. Inside, it reminds me of one of those demountable classrooms they put up at my primary school when the buildings ran out of space. There are a couple of rooms, and toilets down one end–luckily for me, both a men's and a women's. On my first day here, my boss said the women's toilet was all for me as I'm the only woman in the whole Remelt section. 'If you have to go to South Remelt,' he added, 'we'll give you a sign to hang on the toilet door, because they've only got a men's over there.' I'm so glad I haven't had to do that. The thought of hanging a sign on the door that basically says 'there's a girl in here using your loo' is so embarrassing!

I go into one of the rooms, sit down in front of my IBM PC and switch it on. While I'm waiting for it to boot, I notice that I've picked up that damn pen again.

I have a history with this pen. Back on my second day here, I was thinking so hard that I didn't notice I was flicking it until a splodge of ink flew out and landed on my chest, staining my shirt. I was so embarrassed and really upset about the shirt which I bought specially for this job. But my boss is so nice–he just laughed and told me to soak the shirt in meths before washing. He was right and the stain came out, much to my relief. And then, what do you know, the very next day I did the same stupid thing again! You'd think I would have learnt to only use this pen for writing, but here I go again, catching myself flicking it. I wish I knew why I'm always doing such dumb things.

OK, enough of that, the PC is waiting for me now. I type the command to load Lotus 1-2-3, the 'spreadsheet' application that I've been teaching myself these last weeks.

Before I got here, someone had set up this PC and connected it to a thick, black cable which runs through a hole in the wall. At the other end of this cable is a mainframe computer, though I couldn't tell you where it is or what it looks like. All I know is that, through this cable, I'm downloading numbers to load into Lotus 1-2-3.

The numbers tell us about the contents of those giant crucibles on the factory floor. They list the composition of 'the melt', which is what they call the melted down cans and construction waste and other recycled aluminium that's in there. And, as I now know, the melt is not just aluminium, it also contains a long list of other metals and contaminants. Before the melt can be poured out, it must be adjusted so that the percentages of contaminants are within allowed ranges.

'How do you take the contaminants out of the melt?' I asked my boss on my first day.

He shook his head and replied, 'No, we can't take anything out. We have to add pure aluminium or steel to bring the percentages down.'

To get the numbers, samples from the melt are taken and analysed by a metallurgist. Their figures are fed through the mainframe, and the data that comes out is used to work out how much metal should be added. 'Pure metal is expensive,' explained my boss, 'and we don't want to use any more than we have to.'

Currently, the data from the mainframe is printed on a roll of paper, and the foreman must trawl through lines and lines of tiny numbers to work out how to balance the melt. It is a slow and error-prone task, and my job here is to fix that.

I've done a lot in eight weeks. First, I had to understand what the numbers meant, and then I had to figure out how to create the colour-coded graphs and charts that my boss wanted. To do this I had to teach myself how to write macros: short bits of code that fetch the data from the mainframe, add it up, calculate percentages, and produce the charts. I've got it all working at the click of a button now–it's so cool!

There's still a couple more reports my boss wants, but he thinks we're ready, today, to show what I've done to a Very Important Person. I click through all the charts, making sure they display correctly. I remember that I wanted to check the colour-coding on one of them… but no, it looks fine. Phew, I think I'm ready!

My boss comes by and I show him the latest charts. He looks really pleased. He's been so nice, really encouraging. I'm this nineteen-year-old kid who turned up here with no clue about Lotus 1-2-3 or macro programming or how not to make a total mess of my shirt, and yet he's always said he knew I could do it. I wasn't so sure, but I guess he was right!

The Very Important Person arrives to see my work–it's the foreman, clomping into the office in his heavy boots. He greets me with a friendly smile then removes his hard hat and combs back his sweaty hair with his fingers.

My boss takes the lead, describing the purpose of each chart as I load them. The foreman leans forward, looking closely at the screen, not saying much other than the occasional 'Mmhmm'. When I finish, he straightens up and says, 'This'll work.' He replaces his hard hat, gives me a nod, and turns, clomping back out of the office. My boss grins at me and says, 'Well done!' before hurrying out to catch up with the foreman.

I take a moment to enjoy this feeling of intense satisfaction. I have made something that is genuinely useful! The difficult task of interpreting all those numbers has been replaced with charts that can be understood at a glance!

What's more, I have a feeling that my charts will go on being useful for a long time after I return my baby-poo-brown shoes and my hard hat and go back to uni. And that's amazing!

*

I returned to university after my aluminium adventure with a comfortable increase in both my bank balance and my interest in a computing career. Classwork had been theoretical, its purpose often unclear, but this job had been practical and impactful. I had been given a problem, I had figured out how to solve it, and I had seen the very real and positive effects of my work. It was exhilarating!

As I would learn in my next job, these feelings of autonomy and usefulness aren't just a nice-to-have. To me, they are essential.

The sweatshop one

In the second half of 1989, my third and, I thought, final year of university, I submitted my CV to the university's Graduate Fair. It was a day of rapid interviews, speed-dating style, in a big hall. I was selected by a handful of prospective employers and, a week or so later, one of them offered me a job.

I wasn't sure which interview had prompted this offer of a junior programmer position, but that didn't stop me from accepting right away. I had enough stress with a major group project running behind schedule and my final exams coming up, and it was a relief to not also have to worry about how I was going to support myself.

At the beginning of December, after passing all my assessments and exams, and recovering from the end of year partying, I embarked on the next phase of my life. Bright and early on a Monday morning, I caught the train to North Sydney and walked a few blocks to the office building that was my new place of work.

I met first with an executive whom I recognised from my speed-dating interview. He sat me down across from him at his enormous leather-topped desk, while the blue sky of a Sydney summer's day glowed warmly through his floor-to-ceiling windows.

I would, he said, be joining the team of programmers who worked on their warehouse management system, but he told me nothing further about the business. Instead, the main point of the meeting seemed to be a particular clause in my contract.

'I want to make sure you understand that your contract requires you to work forty hours per week, not including breaks,' he said.

I nodded, yes, I had read the contract before signing it. This was my first permanent position, and having nothing to compare it to, I hadn't thought anything about this clause.

He continued, 'To compensate you for not getting a paid break, we buy you lunch. Everyone eats together at the café downstairs.'

'That sounds lovely,' I said, smiling and thinking, 'Well isn't that nice, this is going to be a good place to work.'

Quick introduction over, the exec stood and led me back out of his office. He waved a hand towards an elegantly furnished seating area and a few more offices beyond, backdropped by more floor-to-ceiling windows and a view towards the harbour. 'Executive area', he summarised, before turning and leading me into a corridor at the back of the building.

We passed the toilets and a tiny kitchen before arriving in the programmers' room. Here, unlike the main office, it was dark and cramped with ten desks wedged in as tightly as possible. The only window looked across a narrow lane to the back of another building—not a glimmer of direct sunlight made its way down the steel and concrete crevasse in between. The occupants of the room sat hunched at their terminals, tapping away, barely glancing up to give me a brief nod or half a smile.

A man rose from his seat, introduced himself as the team leader, showed me to an empty desk, and handed me my login details. After I logged in, he gave me a quick briefing.

'So, if you go into your messages here, you will see any tasks I have sent you. Each one has a unique ID number, and before you start on a task, you must go to the command line and type "ton number". "T-on" stands for "task on", right? As soon as you "ton" another task, the old one is automatically switched off. Now here's the most important thing, and I cannot stress this enough: if you take any kind of break, if you get up to go to the bathroom or make a coffee or go to lunch, you must first type "toff" at the command line to turn off any active task. You are required to work forty hours of "task on" time per week. Got that?'

'OK,' I nodded, 'that all sounds straightforward enough.' I took my seat and waited for my first task to arrive so I could get started on the forty hours.

As the junior in the team, my tasks were mind-numbingly trivial. I never got to write a whole function on my own and instead was assigned modifications to existing ones. I had no clue what any of them did or how they fit into the whole. My task description might read 'Change the input variable in function 1392 from a to b'. I'd 'ton' that task, find the function, finish the change in about five minutes, 'toff' it, and then wait for the next one to arrive in my inbox, wondering how in hell I was ever going to make my hours that week.

At midday on the dot, everyone typed 'toff' and trooped downstairs to the ground-floor café. On my first day, as we took our seats around a large table, I noticed that only the programmers were there.

'Where are the people from the other part of the office?' I asked.

'They don't eat with us,' someone replied. 'They come down at 12:30.' Apparently, the corporate class divide didn't just apply inside the office.

Within moments of sitting down, the wait staff dropped plates of food in front of us. We were given no choice and ate whatever they served. The food was fine, but with its speedy arrival, there was very little opportunity for chit-chat and getting to know my co-workers. Everyone tucked in, and within fifteen minutes, we were back at our desks.

By the end of the first month, this lunch deal already felt less like generosity and more like a trap designed to keep me locked in the building all day. Increasingly, I declined the 'free' lunch just so I could get outside into the fresh air for half an hour. The other programmers shook their heads at the profligacy of buying a sandwich instead of eating with them, but I had to. This company was trying to turn me into a coding robot who didn't even get to decide what she'd like to eat.

Worse still was the implacable tyranny of the task logging system. Fridays were interminable as I watched my counter creep up—the second it hit forty hours, I left.

I complained about it to a colleague, but she just shrugged and said that's how it was. She told me of a former employee who was fired after he was caught hacking the task logger and upping his hours. When she mentioned his name, I was surprised to recognise it: he was a former student from my uni, a year ahead of me in Computer Science. In a great scandal, he had been expelled after hacking into the main School server. It sounded like he'd continued with his hacking ways since then, but I was so sick of this job and the task logger that I said, 'Good for him!'

After two months of the claustrophobic programmers' room, the dragging hours, and the mind-numbing tedium, I could take it no longer. I handed in my resignation, effective immediately.

The executive who had hired me was shocked, protesting, 'You're supposed to give us three months' notice!' But I had, I reminded him, read my contract, and it clearly stated that either party could terminate the agreement without notice during the trial period. He had thought this entirely for their benefit while they evaluated me, but it gave me the right to evaluate them too. And my evaluation was that if I stayed there another day I would die of boredom.

After explaining his contract to him, I added, 'I don't even know what the product does! I don't know who the customers are, and I've never seen it being used!' He seemed genuinely surprised that one of the drones in the back room might want to understand what they were building.

*

In light of the more positive work experiences I'd had in the past, particularly the factory one, a different me might have shrugged off this experience and gone looking for another job. Instead, two months in that dark and cramped programmers' room had altogether drained my enthusiasm for a computing career.

My Computer Science degree had prepared me to be a programmer and little else, and as far as I knew, that sweatshop was what being a professional programmer was like. 'If that's what a degree in Computer Science gets you,' I said to myself, 'forget it!'

It would take many more experiences of work before I learnt the truth: it is not the *type* of work you do but *where* you work that is, by far, the most significant factor in job happiness.

But I didn't know that then. I was still only twenty years old, and I had another choice available to me—a far more attractive one. I could delay my transition into the 'real world' by extending my undergraduate degree with an additional year of Honours. I could get back my tutoring job and keep on living inexpensively in my student share house. I needed more time to figure out what I really wanted to do, because whatever it was, it was *not* going to be with computers!

I would, of course, find my way back to computers in time. When I did, with my first IT support and administration job, the lessons I had learnt from those earlier work experiences proved invaluable. The Apple job encouraged me to speak up when I didn't understand something, and to trust my own evaluations over marketing spiel. The tutoring job shaped my approach with my users as I guided them towards identifying and solving their own problems. The factory job inspired me to look for efficiency gains and write my own tools when ready-made ones didn't exist. And after being so constrained in the sweatshop, I relished the autonomy and pace of IT.

We'll get to that IT job later. Before that, I went back to uni, and there was much more I wanted from it than education.

4. How to pay for play

AT THE TIME I QUIT THE PROGRAMMER SWEATSHOP I was living in a five-bedroom student share house. It was a crumbling old terrace, a short walk from the university, with rising damp, a rotating cast of housemates and their friends and lovers, and an eternal pile of filthy dishes in the kitchen sink. And I loved it.

I learnt many things in that house that proved far more useful than the theoretical lessons of school. Sharing a house taught me to be tolerant of other people's differences. Equally cash-strapped flatmates shared their budget hacks, like how to turn the cheapest two-minute noodles into a proper meal. And most importantly, I learnt how to have *fun*.

For a very long time, I'd been no good at fun. As a teen I was always being told by friends, and even sometimes by teachers, 'Carol, don't be so *serious*, it's FUN!' What they didn't know, and what I didn't yet understand, was that my inner Smellbomb was far too vigilant for such frivolities. At primary school she had been taught, too many times, that if she ever let her guard down, someone would make her pay.

Now, after three years of university life, I was letting go a little, and a lot of that was thanks to where I lived. In that house I had learnt to relax on the evenings when my flatmate who worked in a video store brought home the latest movie release, and we all piled onto the couch to watch it together. I had learnt to go with the flow on the nights we stayed up late, watching music videos and talking about anything and everything. And I had learnt to give in to sheer unselfconscious joy whenever my music-loving flatmate put on his Guru Josh twelve-inch, and we all dropped whatever we were doing, ran to his room, and danced.

I had fun in that house, and I wanted more of it. From what I'd seen of the workforce so far, not only was there no fun there, but it also sapped my will to enjoy it anywhere else. I wasn't ready for that grey life; I wasn't ready to give up on all the new things I was discovering as a student.

So back to uni I went. I could have done Honours Computer Science, but I didn't even think about it. I had completed a double-major in Applied Mathematics and I went that way instead.

I chose Maths not just because of my new disdain for programming, but also because my best friend Diana was doing it. I'd met her two years earlier and had been immediately drawn to her confidence and her quick, sometimes quite surreal, wit. Since then, our friendship had developed into one of those ones where you know all the daily minutiae of each other's lives.

Diana was delighted that I'd be joining her, and we resolved to make the most of this bonus final year as undergraduates. We'd go to all the parties, and the free on-campus events, and see if we could talk a couple of residential college boys into taking us to one of their big functions. 'And we should do the band competition!' suggested Diana. It was, we agreed, going to be a fun year.

I had another reason for choosing Maths, a secret reason, one so sensitive I didn't even want to admit it to myself. My father was one of the academics in the School of Applied Mathematics, and really, I went back for him.

After my final expulsion from his house, I had continued to see him around campus, though he rarely sought me out. Sometimes I knocked on his office door and suggested he buy me lunch. We'd go somewhere nearby and I'd listen while he talked about the latest exploits of my little half siblings, whom I missed terribly, and about his wife's pregnancy with their third child. Eventually, I was invited back for strictly curtailed visits to his house. I loved playing with the kids, but my father never showed much interest in me and what I was doing. More than anything, I wanted to regain something of the closeness we'd once had.

In retrospect, I was putting far too much pressure on that fourth year of university. I wanted alternatives to the computing career I now viewed as intolerable, I wanted to rebuild my most significant relationship, I wanted my social horizons to continue to broaden, and I really, really wanted to have fun.

*

The problem with fun though is that it's rarely free.

As agreed, Diana and I never turned down a party invite, and it was only polite for me to contribute to the beer or boxed wine purchases, as I fully intended to contribute to the drinking. Sometimes there were a few joints going round, and some dollars to chuck in if I wanted a toke. I didn't particularly like the way marijuana made me feel I was falling backwards, but I did enjoy the way it made everyone laugh breathlessly at the silliest things. That was fun, so I was in.

In first semester, my flatmates and I threw a big party—the type where the house is packed from the open front door through to the scraggly back yard, the music is turned up, the toilet paper runs out, you're knee deep in empty beer bottles the next day, and everyone declares it a great success.

Before the party, one of my flatmates had asked if I wanted to go in on some speed for the night. A year earlier this suggestion would have terrified me, but I'd shed a lot of my constraints since then. For so long, I had tried being the good girl with the good grades who made the good choices, and what had that got me? Kicked out of home and bullshit career options, that's what. I gave my flatmate $20.

At the party, I swallowed a spoonful of bitter, orange oil containing methamphetamine. Ten minutes later, I was off! I tore around the house, barging into conversations to introduce myself to anyone I didn't know, before racing off again, bursting with too many high spirits to stand still. I danced in the dancing room and disrupted the chill in the chill-out room and bounced from one group to another. It was the most unabashed fun I could ever remember having—and all for twenty bucks!

Don't worry, this experience didn't turn me into an overnight addict. Twenty dollars was still a lot of money for me then, and now I was on the fun train, there were many other things I wanted to try.

Diana and I found dates for a college Informal. This was a huge party with an open bar, a well-known live band, and lots of drunkenness. I had a good time, but on the ratio of fun to expense, it was poor value. I had to buy a new dress and shoes, and when our dates took us out for dinner before the party, they picked a pricey restaurant and expected us to go dutch.

I spent even more on the band competition, but I saw that as an investment. Diana and I had a shared history of learning the violin, but unlike me, she had kept on playing. Knowing that I had a violin in the corner of my bedroom, she suggested we build the band around two fiddles.

'But I haven't played in years,' I warned her.

'You'll do fine,' she said, 'I'll write our parts, and I'll make sure yours is something you can play.'

We recruited one of my flatmates on guitar and one of Diana's on bass. He brought in a singer and a percussionist, and together, we became *Fred the Bison*.

My violin was my father's old one, the one he'd first learnt on as a teenager and had long since replaced. He had given it to me many years before, but it wasn't in playing condition and hadn't even come with a bow. I'd taken it with me when I moved out of his house, with some idea of picking it up again in the future. Well, the future was now. I paid for a service and a new set of strings, and I bought an inexpensive bow and a pickup for amplification. All up, I spent around $700 of my sweatshop earnings—far more than I'd paid for any one thing in my life—but on the fun to expense scale, I considered it excellent value.

I loved being in the band. I loved the creativity as we worked out the songs, covers, and originals for our competition set. I loved the feeling of being part of a group, united by a common purpose. Getting up on stage terrified me, but the natural high afterwards was even better than speed. After we won our first heat, someone threw us an impromptu party. There is a photo of me from that night, dancing in the kitchen, singing 'Come on Eileen' at the top of my lungs. I love that photo for the moment it captures when I'd allowed myself to let go and be unselfconsciously happy.

In the end *Fred the Bison* was a short-lived beast, getting to the quarter finals before bombing out. No one else wanted to carry on with the band after that, but a spark had been lit in me. There was something I needed in that experience of playing in a band, something I would seek out later.

But it would have to wait. As my Honours year progressed, my academic workload became so all-consuming that there was no more time for fun.

*

In addition to coursework, I had to write a fifty-page paper on any area of Applied Mathematics. My paper was titled, 'Modelling the Aggregation Phase in *Dictyostelium discoideum*' which, I'm sure you're aware, is a very weird species of slime mold. It wasn't a topic I could ever have dreamed up myself and had been suggested by my Biological Modelling professor.

Once I started looking into slime molds and learning about their astonishing lives that fall somewhere between plant and animal, I was hooked. I went in deep, reading every relevant book and research paper, and bending my brain around the increasingly complex mathematical models that sought to understand how these single celled organisms spontaneously come together into something that looks and moves like a gelatinous slug.

Every evening for many weeks, I left behind the distractions of my share house and walked through the darkened university campus to the Honours Maths common room where I worked uninterrupted, deep into the night. When my paper was finally done, all neatly typeset and bound and ready to hand in, it was the proudest achievement of my life.

Between the fun and the slime molds and my ongoing tutoring work, I hadn't given any serious thought to a post-university job. I'd had no trouble finding work the previous years and I assumed that, if anything, it would be easier this time round. After all, I'd now have a degree *with Honours*. I put job hunting to the bottom of my to-do list, something to worry about after exams.

And then, late in the academic year I realised, with a sickening drop in the pit of my stomach, that I was in trouble. The sweatshop money had long since run out, and my tutoring work would end before the pre-exam study period. I needed money to get through exams, and I didn't have it.

I asked Student Services for help, and they gave me a $500 hardship loan—but even after cutting all non-essential spending and living on toast and two-minute noodles, it wasn't enough. So, I swallowed every bit of pride I had and asked my father. He lent me a further $500 on the strict condition I repay every cent as soon as possible.

I had wanted to see much more of him that year and, foolishly as it turned out, I had taken his class. There were other electives I could have chosen instead, but my secret goal had driven me towards his. It was never going to end well, but I didn't know that when I signed up. I convinced myself that taking his class was a great idea, right up until the day of the exam.

That morning, as I sat down for the exam, I felt confident. I knew the material so well that I had talked Diana through some of it on the phone the night before.

And then my father dropped an exam paper before me, walked to the front of the room, told us to start, and my mind went completely blank.

I bombed his exam, and for that reason and that reason alone, my expected first-class Honours was downgraded to a second-class. My father and I never spoke about it. For years afterwards, the merest thought of how disgusted he must have been as he marked my sorry excuse for an exam paper was enough to turn my guts into poisonous, writhing snakes.

At the time it never occurred to me, or apparently to anyone else in the faculty, that some sort of special consideration might have been warranted. I blamed myself for failing to accurately predict the future when I took his class, even though there was no way I could have. In the hundreds of tests and exams I had sat in my life, I had never panicked before.

After that disaster, I did what I always did: I pushed the unbearable feelings down into the bedrock of my shame and self-hatred, and I pretended that I could ignore them.

And so it was, with my emotions as threadbare as the clothes I had been patching for the last three years, I got through to the end of exams. And found a recession there waiting for me.

*

In 1990, the Australian economy suffered its worst recession since the Great Depression. I was looking for work in a market where the finance and insurance companies that normally employed maths graduates were laying people off.

I only got one interview, for a position as a junior financial modeller with a small investment fund. The interview seemed to go well, and I was optimistic, but then I heard nothing. I waited for two weeks, and then I phoned them.

'You're the best candidate we've seen,' I was told, 'but we have a couple more interviews to do. Ring again in a week if you haven't heard from us.'

This went on for several more weeks. I'd phone, be given some positive words, asked to hang on a bit longer, and advised to call again next week. I was inexperienced, and I didn't know that if someone really wants you, they'll call you.

Already, the money I had borrowed was gone, and I was using my emergency credit card to pay rent and bills. A flatmate suggested I apply for the dole, but I wouldn't even think about it. My mother lived on a pension, but for me, the idea of accepting welfare felt too shameful. 'That's for people who really need it,' I thought, refusing to accept that I was now one of those people.

As the holiday season approached, I worried incessantly. In Australia, the Christmas-New Year period coincides with the summer holidays, and it's an impossible time to get anything done. Even in a good year you'd be lucky to find a corporate job during that period, let alone during a recession.

I made my final call to the investment fund on the morning of Christmas Eve. For weeks they had prevaricated and delayed, but now they had an answer.

'It's a no from us,' said the man, 'we just don't think you'll fit in'. He gave no reason why they thought this, nor why they couldn't have said so sooner.

I felt sick. I had gambled everything on this job and now here I was, with nothing left but debt. I'd prioritised fun that year, I hadn't been as careful with my money, and now I had to pay for it with my misplaced pride. Resigned, I went to the social services office, and I joined the queue of people applying for the dole.

Later, on the long walk home, I did calculations in my head. I had a couple of weeks before the dole payments started, perhaps longer considering the time of year. Rent was due in early January, and there was no way I'd make it. When I got home, I did the second thing that day I really hadn't wanted to do: I told my flatmates I'd be moving out at the start of the new year.

There was only one place I could go to avoid homelessness. It was a choice I would not have made in any but the most desperate of circumstances.

5. How to rescue yourself

YOU KNOW THE TROPE: THE REDEMPTION ARC OF a person who learns to rescue themself only after hitting rock bottom. Well, this chapter is my version of that story. It's going to get dark but stick with me, because it ends with the decision that led me to IT.

Christmas 1990 wasn't much of a merry one. The day before I'd lost my one hope of a job, applied for the dole, and told my flatmates I would soon be moving out. As usual it was just me and my mother on Christmas Day, and as usual it was a quiet and strained affair. This year felt even more gloomy because I was going to have to ask if I could move in with her.

I've never had what you might call a normal mother-daughter relationship. From her I remembered no affection, no kind words, and the bare minimum of attention. My earliest memories are of her saying, 'You're not the centre of the world!' and 'I don't want anything to do with your games!' I learnt early on that I couldn't depend on her.

My mother had a long-term, undiagnosed mental illness. Before my memories began, she had been a schoolteacher but was pensioned off for life after a series of 'nervous breakdowns'. She defined herself by illness, putting it at the centre of her identity—but despite her history of mental ill health, she maintained that her problems were entirely physical. She just needed to find the right therapist, the right treatment, the right dietary regime or wellness practice, and then she'd be saved.

I suppose at one point she looked to my father for rescue, but he abdicated that role, leaving when I was seven. She then devoted her life to looking for salvation in all the wrong places. She followed one guru after another: homeopaths, acupuncturists, herbalists, allergists, and for a particularly bizarre period in my teens, a Japanese apocalyptic cult that promised divine healing.

Some gurus she rejected quickly, but others she followed obsessively. She spent her limited disposable income on their expensive products and worked herself into a frenzy following their impossibly strict regimes. Often this involved food, and her mental health deteriorated further as she attempted to live on tahini and mung beans. She was bulimic for many years, though I could never get her to admit it.

When my father got himself out, he left me behind like a hostage to his own broken vows. That his leaving coincided with the start of the Smellbomb years is no coincidence. My mother was incapable of looking after me, and from then on, I had to feed, clean, and take care of myself.

I no longer saw her in the mornings, and I got myself up, dressed, breakfasted, and to the school bus on time. It's incredible to me now that I did all that alone from seven years old, but there was only so much I was capable of. When I arrived at school, it was with tangled hair, a crumpled and dirty uniform, and a school bag empty of food for the day. I was neglected and clearly unprotected and I probably did smell—I was, in short, the perfect target for bullies. Other bad things happened under my mother's inattentive watch, but we'll get to that later.

I spent the next four years fervently hoping that my father would rescue me. I was a child, too young to deal with the bullies, with my mother's lack of care, with her periodic descents into bizarre and distressing behaviour, and with the people she allowed into our lives. Children cannot rescue themselves, they need responsible adults for that, but my father was in no hurry. He was far too busy with his unencumbered bachelor life and a series of girlfriends to give me much thought.

Eventually, he agreed I could move in with him for high school, on the condition I got into the selective state school near him. Scoring a place there really did feel like rescue—not just from the primary school bullies I was leaving in my dust, but from living with my mother. I hoped then that I could finally have a normal, loving home.

I had a good year or so with my father, but then he remarried and I found myself back in a whole new bullying situation, again with no rescue in sight. And meanwhile, I still had to spend every weekend with my mother and her obsessions, her paranoïas, her eating disorders, and her periods of untreated psychosis. I stopped hoping for rescue and instead resigned myself to waiting for the one thing that would definitely release me from this crappy childhood: growing up.

By university I had stopped visiting her regularly, but she had a sly way of making her problems my responsibility. She was good at finding nice people to feel sorry for her, and they'd help her out for a while—until she became too much for them, and they'd come looking for me. I'd get phone calls from strangers saying, 'Your mother needs you!' Some even went on to categorically state that I was a Bad Daughter for failing to look after my Poor Mother properly. I was barely an adult and yet the world seemed to think it was my job to rescue her from herself.

I did my best. When the calls came, I dropped everything and made the long trek out to her place to the north of Sydney. I'd find her shaky and paranoid, not having eaten for days, and convinced she was dying because of electromagnetic radiation or mould or the 'tainted DNA from an ancestor who had syphilis'. I knew from experience there was no point trying to get her to a real doctor, so I did the only thing I knew how: I bought food, cooked, made her eat, and stayed for a day or two until she calmed down. And then I fled the craziness and the guilt, back to my own life.

Now, unemployed and without prospects, I had lost that life to flee to.

Unlike me, my mother was very happy I would be moving in with her. She loved the idea of having me there 'to keep me company', while I could think of nothing worse.

*

Back at my share house after Christmas, I boxed up my meagre possessions. Needing a sympathetic ear, I rang the one person I knew who would immediately understand how I felt. It was my old friend, and my mother's former next-door neighbour, Tiffany.

I first met Tiff over my mother's back fence when I was thirteen. We quickly became best friends and after that, I spent most of my weekends at her place. Her parents were nice, and unlike my mother's house there was always food in their fridge to which I was welcome to help myself. And Tiff was great company. We danced around her living room to our favourite pop songs and argued about which member of Duran Duran was the cutest and developed a whole language of in-jokes. I loved her like a sister.

When I told her I was moving to my mother's house, I didn't have to explain a thing.

'Yeah, you don't want to live with her,' said Tiff, 'she'll drive you crazy. But there's an easy solution.'

There is?

'I've just signed a lease on a house in town, not far from where you're living now. There's a spare room, come live with me! You don't have to pay rent, and you can stay as long as you need to.'

It was a Christmas miracle! I'd been rescued!! And it felt particularly miraculous, considering the offer had come from someone who had recently been in no better financial state than I was.

Since quitting high school, Tiffany had drifted from one unskilled job to another. For some time, she had been living with her boyfriend in a caravan, both of them just scraping by. I was aware of the new profession she'd been dabbling in, but not that it had become lucrative enough to rent a whole house.

*

I moved in with Tiff and her boyfriend at the start of January 1991. We didn't have much stuff, and the three-bedroom house echoed with emptiness.

'I'm going furniture shopping,' said Tiff. 'Come along!'

We went to a big department store where I watched her buy two couches and a fancy stereo, paying with cash that she had stuffed in her purse. If the shop assistants found this as strange as I did, they kept it to themselves, counting out the crumpled notes without comment.

Next, we went to a shop that sold evening wear, all taffeta and lace and sparkles. The shop assistant recognised Tiff from a prior visit, and she fluttered around, pointing out new items on the racks, enthusing about how they'd match her pretty eyes, or suit her lovely figure. Tiff beamed, lapping up the attention and the flattery.

Not so long ago, she and I had been a couple of teenagers trying on clothes in shops where the staff completely ignored us, knowing full well we had no money. Tiff was thrilled by the change in her fortunes, and proud of her spending power.

I could see that she also enjoyed having me there as a witness. I had always been the better student, and this had sometimes stoked a little insecurity in her. Now the tables were turned: I was down on my luck, and she was a *success*.

Back at the house, her boyfriend had returned from work. I made dinner for the three of us and we opened bottles of beer and toasted all the great times ahead.

Then, it was time for Tiff to get ready for work. She invited me to her room to hang out while she changed into lingerie and put on one of the new dresses she'd just bought. She packed on make-up and sprayed her wavy, golden-brown hair into place. All ready, she called a taxi, kissed her boyfriend goodbye, and left for the nighttime establishment where she plied her trade.

I watched her go with mixed feelings. I found the thought of doing what she did completely repulsive, but she wasn't me. Since she'd first told me about her new job, she had repeatedly assured me that she was happy with the work. Other than the icky feeling in the pit of my stomach, I could think of no logical reason to disbelieve her. Furthermore, she had made it abundantly clear that my duty as her friend was to give her my full support. Her boyfriend had been given similar instructions. He was trying.

At the end of each night, Tiffany was paid with a wad of cash which she tossed into a cupboard by her bed. Never having had much money, she had no idea how to manage it. None of it went into a bank account, let alone investments or a superannuation fund. In Australia, her job was not illegal, but failing to declare and pay taxes on her income was. She didn't even think about it. She was a kid, only recently turned twenty-one. She splashed the cash on whatever took her fancy, and she didn't think about the future.

I had always heard that drugs led to prostitution, but in that house, I witnessed the reverse. Tiff and her boyfriend had never had the money for more than an occasional indulgence, but now, anything they wanted was within reach. And what they wanted was a steady supply of speed and ecstasy and marijuana.

At first it was fun, every night a party. The boyfriend and I, in our capacity as Tiff's support crew, started the evening by sharing a couple of lines of speed with her, just to get her pepped for work. Then Tiff called a taxi and departed, leaving us full of deranged energy and looking for something silly to do. One night we competitively did sit-ups, thrilled by how effortless it felt. For the next couple of days, each time we'd spot the other wincing from the strained muscles, we'd laugh, which was even more agonising, making us laugh even more, until we were doubled over, helpless with mirth while our eyes streamed from the pain.

The boyfriend befriended a pair of brothers who lived up the street, and they soon became regular visitors to the house. The brothers were obsessed with new music—Madchester, shoegaze, and grunge—and often arrived with albums they'd just bought. We'd get some chemicals into us and stay up all night playing records. In one session, I lay on the floor, a stereo speaker on either side of my head, blasting the Happy Mondays directly into my ecstasy-filled brain, my body coursing with endorphins, feeling like a god.

Tiff usually returned between 3 and 4 am. She never seemed bothered by the drug parties she was funding, and instead she'd be pleased to see us all wide-eyed and awake. She'd have a few lines, or half an e, and then regale us with tales of her night, while the boyfriend sat silently by with a bowl of weed, packing one bong cone after another.

*

The weeks drifted by. After years of study and work, there was nothing I had to do other than go to the dole office every two weeks and explain why I hadn't found a job yet. I was still looking, and applying for anything that specified a Maths degree, but because of the recession there wasn't much, and I didn't get a single interview.

I felt untethered, cut off from the life I'd lived before. Without the structure of study and work to prop me up I felt as amorphous as a jellyfish, listlessly drifting, washed along on the ripples of other people's lives. I didn't see any other friends, not even Diana. I was certain that everyone was getting on with their lives, finding jobs, being happy. I felt too much of a failure to reach out to any of them.

I didn't call my father either. I was deeply ashamed about failing his exam and about my inability to pay back the money I'd borrowed from him, but there was something else stopping me too: I was angry. All those years when I needed him to step up, to be a responsible parent, he'd completely failed—and he didn't even have the decency to admit it.

But even as I directed my anger towards him, I turned more on myself. I had worshipped him, believing implicitly that he was always right, even when anyone could see I was being bullied and neglected. What an idiot I had been. I despised myself for having been a vulnerable child who had loved and needed her only functional parent.

I had a lot to be sad about, so I don't blame my twenty-one-year-old self for welcoming the distraction of the drug parties. At first it felt like a holiday, just a brief departure from the real life that I would soon return to. January passed in this way, and we were well into February before I realised that the other members of my little world did not see things the way I did. They didn't want this to be a holiday—they wanted it to be their real life.

Tiff's boyfriend started on the drugs earlier each day. When we'd first moved in, he'd had a job, but he'd quit soon after, saying that he didn't see the point of working for peanuts when Tiff was bringing in so much. I suspected he was silently adding 'from screwing other men', but if so, he never admitted out loud to being anything other than 'cool' with it.

With little else to do, he'd have a line of speed with a bong chaser at any time of the day or night, and his previously gentle personality became volatile, even aggressive. When I tried to talk to Tiff about it, she said he was fine. And so was she. They were both fine, it was all fine, *OK?*

The brothers from up the street were also going off the rails. The younger of the two, only seventeen, was so enamoured with Kurt Cobain and Perry Farrell that he thought drug addiction was a valid life goal. He wasn't going to pass up such ready access to free narcotics in his quest to emulate his heroes. The older brother, previously the responsible one, was nursing a broken heart and throwing caution to the wind. He was on the verge of losing his job because he kept missing work after our long nights.

On one of those nights, I had a heated debate with the older brother about drug use.

'The only reason to do drugs is *fun*, like a roller-coaster ride,' I argued.

'No, no,' he insisted, 'there's lots of good reasons. Like giving yourself a boost or calming yourself down, getting to sleep, or, like, having a great trip where you really learn something about yourself...'

I shook my head at him. I didn't have the words to explain exactly why I was so sure, but somehow the self-preserving part of me, the part that had got me through some truly awful childhood experiences, knew that to assign purpose and function to drugs is to give them power over you.

There is a sad addendum to that conversation. A couple of years later I ran into the older brother in the street and was shocked by how haggard he looked, a shadow of his former self. 'You know that argument we had that night?' he said, 'I wish I'd listened to you. You were right.' I felt no triumph.

*

The feeling that Tiff had rescued me with her offer of free accommodation was dissipating fast. I was worried about her, and the boyfriend, and the brothers, and myself. I knew I needed to get out of this place, but the only thing that would do that was an income.

While scouring the newspaper's employment pages, I had seen listings for Computer Science grads, but I'd skipped past them. I was still haunted by the claustrophobic memories of the programmer sweatshop, and I was still stuck on the idea that computing was not for me.

But I had to acknowledge that maths wasn't getting me anywhere. I tried widening the net, but in that time of high unemployment, even listings for low-paid service jobs demanded a couple of years' experience. I managed to score an interview to become an usher at the Sydney Opera House, but when I arrived, my heart sank on seeing the crowd of other young hopefuls. The waiting room looked like one of those casting calls you see in movies, where there's a sea of people who don't look so different to you, all going for the same part.

My mood sank even lower as a particular day approached that I was dreading. My graduation ceremony was scheduled for early March, and I wasn't going to be there.

I still owed that $500 to Student Services and the debt prevented me from graduating. Tiff would have given me the money had I asked, but I didn't. My father could easily have helped—as I found out much later, the faculty would have paid the debt had he only mentioned it—but despite knowing exactly when the ceremony was, he never reached out. I didn't expect him to. I no longer thought that anyone would rescue me, because I didn't believe I deserved it.

Graduation Day arrived and I stayed in my room, paralysed by my own hopelessness. I didn't brood, I didn't torture myself by imagining how happy the rest of my Honours class must have been, celebrating their special day with families who actually loved them, I didn't feel angry or sorry for myself. I just existed.

That night drugs were offered as usual, and I took them as usual. Tiff wasn't working and had invited some friends to join the party. The music was turned up, and the room was hazy with smoke and laughter.

For a while, I convinced myself that I was fine and having a great time—but as the high wore off, I slid helplessly into a pit of utter aloneness. The people in the room suddenly looked tiny and remote, like I was watching them on a TV screen at the far end of a long tunnel. I felt like an alien who could never belong in this world. It was unbearable.

Quietly, I got up and left the house. I walked through the pre-dawn dark with no destination in mind, my only goal to put as much distance as possible between myself and what I had become. As the sky began to lighten, I found myself in the grand pavilion of Sydney's Central Station. I looked at the departure boards and saw there was a train leaving for the Royal National Park in ten minutes. I bought a ticket and hopped on board.

An hour later, I was walking down a track in the bush, breathing the eucalyptus-scented air while cicadas screamed their promise of a hot day. By then, I was a veteran of enough drug-fuelled nights to always carry sunglasses to shade my bloodshot eyes, but in every other way, I was completely unprepared for an impromptu hike. I had no food or water. I was wearing my worn-out Doc Martins, my patched old 501 jeans, a little black singlet top, and a cropped leopard-print jacket that I'd found in a charity shop. My hair was in a punky cut I'd done myself, the back and sides shaved and what was left bleached platinum blonde. I must have looked ridiculously out of place. As I went, I crossed paths with a few other walkers on the trail, and I nodded at them in the friendly way you do in the bush. All returned expressions of alarm and hurried quickly by.

Eventually, I found my way to a waterhole, sunk into a sandstone shelf and surrounded by shady trees. I drank clear water from the stream and sat there for a long time. I daydreamed about finding a little cave that I could live in and never having to face the world again. I lay on my back and watched the light playing through the leaves of the trees. After a while, I fell asleep.

A few hours later, I woke from a dreamless sleep feeling more at peace than I had in months. I had been lost, but those hours amongst the trees had allowed me to find my way back to hope. I understood that no friend or parent or employer could rescue me, just as I couldn't rescue anyone else from their own choices. When the danger comes from inside, the only person who can rescue you is yourself.

As I caught the train back to town, I mapped out a plan. First, no more drugs. And second, it was time to start applying for computing jobs.

6. How to find your career

AFTER THAT MOMENT OF CLARITY IN THE ROYAL National Park, I started applying for computing jobs, and my luck quickly changed. I found a position, and it was one that seemed to have been made just for me.

It was in the School of Electrical Engineering at a different university to the one I'd studied at. In one of those split budget compromises you find in cash-strapped institutions like public education, the job was to be half research assistance and half PC support. Ideally for me, both maths and computing were wanted.

I started my new job on the first of April 1991, but it was no April Fool's Day joke. While this job paid less than the programmer sweatshop, it would prove a far better fit. On that first day, all I hoped for was a new beginning, but I would find so much more: I would find my career.

My new office was in the university's imposingly brutalist Tower Building. I took the lift to the twentieth floor and knocked on a door that was opened by a man named

Geoff—someone who was going to become a huge part of my life as my colleague, mentor, and friend. He was dressed very informally in shorts, t-shirt, and no shoes, and his raggedy beard framed a welcoming grin. In the five years we worked together I rarely saw him attired any other way.

Geoff had been one of my interviewers, though from the other side of the table, I hadn't noticed either the shorts or the bare feet. I felt the interview had gone very well, but then, as I left the room, I crossed paths with the next candidate waiting to go in. As I took in his neat suit, shiny shoes and important looking leather organiser tucked under one arm, my heart sank. Obviously, they would give the job to this go-getter, and not to someone as scruffy as me.

So it had been a complete, and very welcome, surprise when I got the call saying my application had been successful. Eventually I would learn that Geoff's vote had swung the decision in my favour. I'm sure he had other reasons, but perhaps on some level, he was looking out for a fellow scruff.

After welcoming me in, Geoff gave me a quick tour of the office. It consisted of two large rooms filled with desks that were occupied by postgraduate students, and a computer bay with a half dozen Unix terminals. In another corner, there was a decent sized kitchen from where you could just about make out the city streets far below through windows that, according to Geoff, had not been cleaned since the Tower was built in the 1970s.

Geoff's desk was near the kitchen, and beside it was one for me. We sat down and he told me a little about his background.

A decade before, as an Electrical Engineering student, he had become more interested in the School's server than his studies. He'd figured out how to hack it from the student lab and crash it at will, and he was clearly still proud about this.

'The best time to crash it was just before lunch because the other students gave up and left, and I had it all to myself when it booted back up,' he said with a devilish grin.

Eventually he was rumbled, but the School's response was pragmatic. Instead of kicking him out, as my old school had done with the hacker whose steps I had unwittingly followed through the programmer sweatshop, they offered Geoff a job.

'It was fine when it was just Unix servers and terminals,' said Geoff, 'but the problem now is all these PCs showing up around here. The academics buy them with their research budgets and then bug me about them. It's got too much for me alone—I haven't been able to take a holiday in three years! So the plan is, you take over anything to do with staff PCs.'

Mission explained, Geoff put me to work. He led me into a small room and pointed at an extra-large PC wedged under a desk. 'This will be the PC network server,' he said. 'Your first job is to install NetWare on it.' And then he left me there, scratching my head over words I'd never heard in that context. Like, what was *net wear*, and how would one go about *installing* it?

It was a crash course in jumping in and figuring it out. Two days, forty floppy disks, and a pile of manuals later, I had the server up and running. Servers need a name and Geoff said I could pick one myself. I chose 'Wildthing', not after the song, but after a picture book I had loved as a child. I was barely out of childhood anyway, only twenty-one. I have a photo that someone took of me that year, sitting at my desk in the office, and I look about sixteen. Even though

I remember being that Carol, I can barely fathom how much she'd already done by then.

After I finished setting up Wildthing, Geoff asked me to help him with a big project. The School of Electrical Engineering was spread over six floors of the Tower, and he was building a network to connect all its staff offices, common areas, and labs. I helped Geoff run coaxial cables through the cavities behind the walls, and he showed me how to terminate them with special connectors that could plug into another cable, a router, or a computer.

Being in a university, we had access to AARNET, the Australian research network, founded only a couple of years earlier in 1989. Through AARNET, we could exchange emails, files, and opinions with people on the research and military networks in other countries—it was the internet long before it was called that. Geoff and I were bringing a networked world right to our users' desktops, and I found that idea very compelling.

Once the cabling was in, the staff PCs had to be modified to use the network. Geoff walked me through the process a few times, and then he left the rest of the PC networking to me. Over the next few months, in a way I'm about to describe, I visited every office in the School to transform previously standalone computers into nodes on a global network.

THURSDAY 16TH MAY 1991, SYDNEY

I juggle everything I'm carrying into one hand so I can knock on the office door. 'Come in' sounds through it, and I open the door and address the man sitting at the desk.

'Hi, I'm Carol. I'm here to put your computer on the network.'

'Great!' he says, and waves me towards a PC sitting on a side table, while he goes back to the pile of papers on his desk.

I stack my bits and pieces on the edge of the table and manoeuvre the PC until I can get at the back of it, unplug all the cables, and shift the monitor and keyboard out of my way. I take my Philips Head screwdriver out of my back pocket and remove the screws at the back of the computer before sliding off its casing.

I have to be very careful about static electricity, so I wrap one end of a grounding strap around my wrist and attach the other end to the computer's metal frame, equalising our levels of static charge.

I remove the screw from one of the metal slot covers and slide it out. Next, I prise open the sandwich-sized plastic box I've brought with me and remove a circuit board, careful to hold it by its metal bracket, not touching any of the static-sensitive components. I slide the board into the empty slot and gently press until its shiny gold teeth lock into place. I screw in the bracket, then replace the PC casing.

I re-plug the power, monitor, keyboard, and something new as well. The circuit board I just installed is a 'network card', and into it I plug a length of coaxial cable.

I had attached the connectors to either end of this cable myself. In my first week here, Geoff taught me how to peel back the layers of insulating sleeve, metal mesh, and firm plastic tubing, attach a connector to the copper core, then fold the layers back in just the right way, before squeezing the connector firmly into place with a thing called a crimping tool. My first attempts were wonky and loose, but I'm getting better at it now. There's a bit of mesh protruding under the edges of these connectors that I crimped, not quite as tidy as Geoff's, but they'll work, and that's the important thing.

At floor level, a small square has been cut into the skirting board and through it, a shiny metal connector protrudes. I crouch down and link my cable to it, enjoying the way the silver plugs lock into place with a satisfying click.

I am now ready to boot up the computer and finish the job. Using my floppy disk of handy utilities, I install a 'driver', a piece of software that tells the computer how to use the network card. I configure the network settings, and then I install the 'NetWare client' which will allow the PC to talk to Wildthing, my new file and print server.

With the computer now on the network, I hand the man a piece of paper with his username and password, and another with a list of handy commands. I show him how to login to Wildthing and get into his personal home directory.

'If you save your files in here,' I tell him, 'they'll be backed up. Also, you'll be able to get to them from the computers in the printer bay—you won't have to take files on a floppy disk anymore. If you need to share a file with someone else in the department, there's a shared directory here that everyone can see.'

The man thanks me and I leave him to it. As I depart his office, my head is high and there's a confident spring in my step. A few weeks ago, I had no idea how to crimp network cables or install a circuit board, but already I can do that and much more. This job is brilliant!

*

When I started the job, Geoff was keen to refer all PC support questions to me—but for the first few months, I needed a lot of his help to resolve them.

'I thought you had a computer science degree,' he'd sometimes grumble after I'd asked him the eleventh question that day.

'Yeah but they didn't teach me any of this stuff!' I replied.

The degree had trained me to become a programmer, or perhaps a chip designer, but they never taught me how to build a server or attach computers to a network or upgrade the RAM in a PC. Like that Apple trainer of years past, my degree hadn't even taught me what RAM, or Random Access Memory, was even for.

It was Geoff who taught me all that, and much more besides. And the more he showed me about the nuts and bolts of computing, the more I loved it. This wasn't some contrived educational exercise, and it wasn't writing tiny bits of code for a whole I couldn't see. This was *practical* work with results that were gratifyingly clear and immediate.

Progressively, over the first year, I built out my PC network to include file sharing, printing, email, and standard office applications. My work made my users' PCs a whole lot more useful, and in consequence, they used them a whole lot more. It became the norm for every staff member to have their own computer, and by the end of the year I was flat out building new PCs, upgrading old ones, fixing problems, installing applications, creating user accounts, managing file permissions, setting up something new on Wildthing...

And at the same time I was, you may recall, also supposed to be a research assistant.

I did one major research assistance job during my first months in the School of Electrical Engineering. A professor had written a book on Global Positioning Systems (GPS) and needed my help to get it ready for publication. I called on the experience I had with maths software from my Honours year to create neat graphs from his hand-drawn ones, and to typeset all his equations. The GPS concept was interesting—it was the first time I'd heard of it—and it was nice to see my name in the Acknowledgements when the book was published, but the task had essentially been to polish someone else's work. This didn't grab me nearly so much as what I was doing on the PC side of my job.

Later, a couple of other small research projects came my way, but I was so busy with the demands of my growing PC network that I struggled to find enough time or attention for them. Towards the end of my first year, I made an appointment to discuss this with the Head of School.

'I'm not a research assistant at all,' I told him. 'There is far too much to do with the staff computers, and you really need me on it full time.'

He listened and he agreed that I made a good point.

'Rewrite your job description with what you're actually doing,' he said, 'and we'll send it to HR and see what they say.'

A few weeks later my rewritten job description came back from HR with the title 'Information Technology Systems Administrator' and an $11,000 pay rise. Eleven. Thousand. DOLLARS!! It was a forty percent increase on my research assistant pay!

I was dumbfounded. Any doubts I'd had about a career in computing were well and truly put to rest. Not only was this IT caper interesting and fun, but it paid well too!

I was twenty-two years old, and I had found my career.

7. How to Me Too

THERE IS A PARTICULAR ELEPHANT IN THE ROOM when the conversation turns to a woman's experience in male-dominated spaces, and for reasons that will become clear in the next chapter, I want to address it sooner rather than later. Many people assume that the incidence of sexual harassment and assault of women is comparably high in IT as a direct consequence of the gender imbalance.

I want to state clearly that this has not been my experience. However, I should also add that it would have been a damn foolish man who tried.

One time only have I been sexually harassed in a place of work, and it was back when I was a university student. The incident occurred towards the end of third year, on the morning of one of my Computer Science exams.

I had been restless at home, too jumpy with pre-exam jitters to relax. So I went in early, heading straight for the Comp Sci building, hoping to run into someone I knew who could offer friendly distraction.

I was just leaving the student computing lab after finding it deserted, when I ran into one of my tutors. Outside the classroom, my only prior interactions with him had been when I'd stopped to pat the elderly blue heeler he sometimes brought into work.

He asked what I was doing there so early, and I explained about the jitters. He mentioned that he had a coffee machine in his office and asked if I'd like to join him for a cup. I thought nothing of accepting. Perhaps I should have worried when he closed his door after showing me in, but most of the staff kept their office doors closed, so I didn't.

He made the coffee, and we chatted about something banal until, out of nowhere, he revealed that he would be marking my exam.

My nerves, temporarily calmed by the distraction of idle conversation, came flooding back. Surely, I thought, there was some academic propriety rule being broken here—but if that was the case, why had he invited me?

Distracted by this thought, I was completely unprepared for what happened next. He lurched at me, grabbing me in a bear hug. He put his mouth right to my ear, his hot, slavering breath moist across my cringing skin, and he said, 'I find you very attractive.'

Gagging with revulsion, I extricated myself from his grip, grabbed my bag, and fled his office. Later, in the exam, I had to force myself to concentrate while part of my brain shouted, 'YOU JUST REJECTED THE MAN WHO IS MARKING THIS!'

I got through the exam, and he marked me fairly, though I'm sure I would have done better if part of my brain hadn't been so busy freaking out. Afterwards, I avoided him like the plague. I was furious with him, but I never made a complaint. I assumed that I'd either be disbelieved or blamed for having gone to his office in the first place—and I don't think I was wrong about that. It was 1989, and there's every chance that's exactly what would have happened.

Also, I was furious with myself. I knew plenty about sexual harassment and assault by then, and I kicked myself for not recognising the signs. Surely, I thought, I should have developed some kind of creep radar by now!

And now, a warning. The paragraphs below the next divider concern child sexual assault and abuse. I have gone forward and back on whether and how much of this awful part of my life to include in this book, but I can't escape the feeling that it is relevant. It is relevant to how I've shown up in all parts of my life, including work. It is relevant to the way I have conducted myself and some of the choices I have made. And it is relevant for *a lot* of other people, some of whom may be reading this book. But unless we talk about it, unless we bring our shame out into the clear light of day and redirect it where it rightfully belongs, how can we, as a society, ever end child sexual abuse? For these reasons, I am choosing to stand up right here and say Me Too.

However, I completely understand if you need to skip forward, and the dividers are there for you. Please look after yourself first and foremost, and know that you are not alone.

*

After my parents separated when I was seven, I lived with my mother in the suburbs. At first, when I visited my father in the city, he drove forty-five minutes each way to pick me up or drop me off from school. After I turned eight, he judged me old enough to catch the train. On my own.

My morning commute to the suburbs went against the general commuter flow, and I was often alone in the carriage. Only a matter of weeks into this new train routine, I was sexually assaulted for the first time.

While I wasn't directly touched, the assault was very graphic and confronting. It was also the most terrifying thing that had ever happened to me. My body completely froze, and no matter how hard I willed my legs to lift me up and carry me away, they wouldn't move. Eventually, the man finished what he was doing and went away, leaving his disgusting discharge all over the floor. It was only then that my body came back into my control.

Shaky and distressed, I moved to stand near the door of the carriage. When I got to my stop, I ran all the way to school. There was no-one there I trusted enough to tell, so I held it in all day. The second I got home that afternoon I blurted out the story to my mother.

True to form, she said, 'I don't want to know about it, tell your father,' and she walked off. So, I phoned my father at his office at the university and I told him what had happened. He listened, and then in his implacably reasonable tone, he said 'What were you doing sitting in an empty carriage? *What did you expect to happen?*'

I was eight years old and had just been told by one parent that the terrifying sexual assault I had suffered was of no interest to her, and by the other than the incident had been my fault. Neither of them reported the crime, my father continued putting me on the train, and he never asked if anything like that happened again. Which of course it bloody well did. I always moved carriages if the one I was in emptied, but the perverts still found me.

A few days past my eleventh birthday, one came to find me in a place where I really should have been safe. I was in my own home when I was raped by my mother's boyfriend.

She'd been dating him for a while, but it was the first time I'd been left alone with him. My mother had travelled to New Zealand for her mother's funeral, and I had been staying with my father. I guess he couldn't wait to get rid of me because the evening before her return to Sydney he drove me out to her house and handed me over to her boyfriend, a man that he didn't know, for the night.

That paedophile took full advantage of my childish trust and ignorance. Only afterwards did I understand something of what had happened, and I never let him touch me again, but that didn't stop the abuse. He frequently exposed himself to me, or he followed me around, offering me money in exchange for acts that he described in explicit detail. Usually, I could evade him easily enough, but there were a couple of terrifying occasions when he cornered me, and then I kicked and screamed as hard as I could until he backed off.

At first, I didn't tell my parents. Considering their responses following the train incident, and from many other things besides, I had every reason to believe that telling them would only make the situation worse. I think I am entirely justified in saying that my own parents *groomed* me for sexual assault. They made it painfully clear that they didn't want to know, and so I didn't tell them.

But after one of the cornering incidents, I was frightened enough to try and tell my mother. As we have already established, she wasn't much of a listener, and she shut me down with the simple expedient of talking over me until I gave up.

I told my father about the abuse when I was thirteen, after it had been going on for two years. We were in Paris—he was there for work and had taken me along. I'd had a fantastic time with him in France, but it was fast coming to an end, and he was about to send me back to my mother while he stayed on in Paris for another six months. With my departure date looming, I broke down in tears and told him I didn't want to go back to Australia without him. I even told him why, saying that my mother's boyfriend was 'always trying to get me into bed with him'.

This should have invoked some sort of protective paternal instinct right? *Right??* But no. My prince of a father again diverted his adult responsibility onto my young shoulders. It was *my* job, he said, to tell my mother what was going on. He put me on that plane, and he sent me back to that situation without any further discussion.

It was another year before I managed to get my mother to shut up and listen to me for long enough to tell her what had been happening. Until then, the paedophile continued to have access to me, and my father continued to allow it. Even after my mother finally broke off the relationship, neither of them reported him, and neither of them ever asked me if I was OK.

So perhaps you can see why I assumed that I'd be disbelieved or blamed about the tutor. And perhaps you can see why I also blamed myself: it was what my own parents had taught me to do.

*

After the tutor incident, I realised I'd made a terrible mistake. An assumption, planted in my brain during my powerless childhood, had lulled me into a false sense of security. Despite knowing full well that women were sexually assaulted, despite the 'Reclaim the Night' marches I'd been on and the slogans I'd shouted, I still harboured a childish assumption that I'd be free from sexual predators once I grew up.

The tutor had just shown me how wrong I was. It scared me to think how much worse the situation could have been—he was a big man and could easily have overpowered me—so I took the incident as a warning. I needed to find ways to reduce the risk of this happening again.

I signed up for women's self-defence classes, taking a couple at uni, and later, when I was working and had the money for it, enrolling in a longer course at a martial arts school. Only a week or two into that course, I used one of my newly acquired self-defence techniques on a real creep out in the real world.

I was walking down a quiet street in broad daylight, when a man asked me for directions. I told him the directions, which just happened to be the way I was going. He fell into step beside me and started chatting. I was wary, but I gave him the chance to prove he wasn't a dickhead. He completely failed on that front when he enquired, 'Are you wearing a bra?'

The first technique I'd learnt was Verbal Self-Defence, and we were taught to deploy it at the very first sign of creepiness. Immediately I turned on him, snatching my sunglasses from my face and waving them above my head to make myself look bigger, with the added advantage that he could see my eyes blazing with fury. I repeated, loudly and insistently, 'I DON'T want you to walk with me I DON'T want you to walk with me I DON'T WANT YOU TO WALK WITH ME!!'

And, like the pathetic weasel he was, he turned tail and fled.

Boy did that feel good!

I understood then that sexual predators are cowards. They went for me when I was a child because they knew I was weak. The tutor made his clumsy pass because he thought the exam put him in a position of power over me. And that loser in the street was testing my female social conditioning to see how far he could push it. The solution was to let those assholes know, early and clearly, that I was not an easy mark.

In self-defence classes I learnt how to ram a man's nose through his face, how to elbow him hard in the throat, and many ways to kick him in his fucking balls. Fortunately, I never had to do any of that, but I think there's something about knowing you could that gives you an unmistakable 'Don't mess with me' air. After seeing off that dickhead with my Verbal Self-Defence, I was never sexually harassed again.

I started out this chapter by stating I had never been harassed in IT, but 'it would have been a damn foolish man who tried'. However, I don't think that's just because of me and whatever I may have been projecting. I genuinely believe that the vast majority of men I have worked with in IT would never dream of making unwanted sexual advances on a female colleague. I have also consulted the available statistics, and while this doesn't excuse any level of sexual harassment, IT workplaces are no worse than average, and a great deal better than retail and hospitality.

If there's one thing the Me Too movement showed us, it is that sexual harassment and assault are rife throughout all sectors and spaces within society. I don't know how *anyone* is tolerating this, I don't know why *everyone* isn't out in the streets shouting, 'Enough is Enough!' Even if you're someone who is lucky enough to never have directly experienced sexual assault or abuse, you're indirectly paying the cost. Survivors are far more likely to suffer debilitating physical and mental health conditions, are far more likely to need financial assistance from the state, and are far less likely to be in stable employment. Surely even the most hardened and unempathetic capitalist should be able to see that continuing to allow one segment of society to use and abuse another for nothing more than a sick power trip *doesn't make economic sense.*

Childhood abuse in particular is the catastrophe that keeps on giving. At twenty-two, I thought I was doing fine with my self-defence classes and my vanquishing of that weasel in the street, I thought I was 'over it'. Really, I hadn't even started to deal with it, and I know now that I can never be fully 'over' such childhood abuse. And yet I have bucked the trend: I have kept my health so far, and I have consistently worked and paid a lot of taxes.

So, what's my secret? Well, it's nothing easy, no 'one weird trick'. Instead, it's been a long and deliberate journey of recovery that began not long after my official elevation to IT Systems Administrator at twenty-two years old. I feel very fortunate to have had the opportunity to work on my recovery at an age when my brain was still developing, and I owe my first steps on that journey to something I stumbled across while working in IT.

8. How to troubleshoot

TO BE A TECHIE IS TO SPEND HALF YOUR LIFE troubleshooting.

Things go wrong, we all know that. Even in the most well-designed and well-managed systems, things break. And who you gonna call? Bug BUSTERS! (AKA your nearest available IT techie.)

Sometimes it's an easy fix—we've seen it before and we know exactly what to do—but often it's a new one, and then we troubleshoot.

The worst thing you can ask us at this point is, 'And how long will this take?' We don't know, we're searching for a needle in a haystack here. Sometimes we'll luck out and find it straight away, but sometimes we're in for a slog, and you're just going to have to be patient.

Even if we go quiet on you, understand that we're picking through log files, googling error messages, triple-checking patching and standard settings, trawling online forums, roping in colleagues to help, running network traces, escalating to vendors, and experimenting with triggering the fault in a slightly different way to see if it changes anything. If we've still got nothing, we start all over again, hoping to spot something different this time around.

I always hate that moment when I'm confronted with a fault I can't quickly resolve. 'Not this a-bloody-gain,' I sigh to myself. 'I don't have the time or the energy right now. Why can't things just work!!'

But once I'm engaged with the bug hunt, it can be intriguing, particularly when there's no one standing over my shoulder shouting 'AND HOW LONG WILL THIS TAKE?!' The neat bonus of troubleshooting is that you learn far more about how things really work when they don't. You only truly grow your tech expertise by fixing things when they're broken.

Of course it was Geoff who taught me how to troubleshoot effectively. 'Only change one thing at a time, and then re-test,' he advised. 'And don't be content with just stopping the fault—figure out what caused it. That way, you'll know how to prevent it from happening again.' I have been faithfully following his troubleshooting guidelines ever since.

Geoff also introduced me to an online place where I could ask other people my questions, instead of bugging him all the time. It was a system of message boards called Usenet, and back then in the early nineties, it was the pre-internet's first social network. Think Reddit, but without the pictures, advertising, keywords, or search.

My first forays into Usenet were to 'comp.sys.novell' where system administrators swapped advice and troubleshooting tips on Novell software products, including the NetWare operating system that I'd installed on my server

Wildthing. I had known nothing about NetWare when I started working with it, and even though the manuals were excellent, they only told me what happened when everything went right. On comp.sys.novell I got the other side of the story: the messy reality of things going wrong.

Usenet demographics skewed heavily towards technical and academic folks, which perhaps explained the general impatience with newbies. I'd been around highly intelligent people long enough to be familiar with that type of intellectual prickliness, so it didn't bother me. People who have put in the hard work to figure something out for themselves can bristle at the suggestion that others want an easy ride on their coat tails. I should know, because I can be like that myself.

On comp.sys.novell, I only needed to be told once to RTFM (Read The (ahem) *Fine* Manual) to ensure that I'd explored all options on my own before asking for advice. If I did ask, I first demonstrated that I was no timewaster by clearly describing the problem, then listing everything I had already tried. And when I asked for help that way, I found that, just like Geoff after he'd got his grumbles about my useless degree out of the way, the netizens of Usenet were exceptionally generous with their knowledge and advice. Usenet became my greatest troubleshooting tool.

But here's another thing about troubleshooting: you can't even get started on looking for a solution if you haven't yet acknowledged there's a problem.

*

Asking for help was, for me, a tricky proposition. Asking for *information* was easy, but asking for *help* was fraught with danger. This is, I believe, common among people who had to be abnormally self-sufficient from a young age. At only five or six years old, I had already been told often enough that 'you know where the kitchen / bandaid / bucket is' that I stopped asking for help when I was ravenous, or when I sliced my finger cutting bread, or when I was throwing up because of the days-old chicken I found in the back of the fridge and put in a sandwich. I had to figure out how to deal with all this, and much more, on my own, from far too young an age.

In my childhood, should I cause inconvenience to my parents, their responses ranged from scorn to blame to occasional bouts of explosive rage. It seemed that the higher my need, the higher the risk of asking for help. Experiences like the one when I, only eight years old, was blamed for a terrifying sexual assault taught me that it was safer to deal with my own needs as quietly as possible.

At twenty-two years old, a year into working at the School of Electrical Engineering, I hadn't even begun to contend with the horrors of my childhood. About the sexual abuse I had long claimed, to myself and others, that I was 'over' it. Logically, I knew none of it had been my fault, and I thought that was all there was to recovery. But the long-term damage done to a child's developing brain by such abuse can't simply be logicked away.

This damage was soon to hit me, just when my life was coming together. Perhaps it was because I finally stopped running, because I arrived at a point where things were falling into place. I thought it would be plain sailing from there on, but instead I had given the trauma a chance to catch up.

My premature sense of safety came from both my steady job, and my life outside work. I'd moved out of Tiff's house as soon as I got my first pay cheque but I couldn't afford much, and my next house had been equally awful, not least because of the raver dude dealing drugs out of the front room. I'd got out of that situation, and now I was renting a house with a couple of girlfriends, Tania and Deb, whom I'd known for years. It was a fun household, with music always playing on the stereo and friends popping in and out. It felt like a proper home, living with people I loved.

After work, I went out with friends and drinking buddies to see bands playing. One buddy had turned into a boyfriend, and while I sometimes worried about his enthusiasm for the drinking, I loved his quirky humour and his easy charm and how passionately he'd argue about anything from politics to metal bands to the relative merits of the French impressionists.

I had also recaptured something I'd wanted when *Fred the Bison* folded after the university band competition. It had taken several months of asking around, but eventually I met a man called Lobster who wanted a violinist for his new band *Unseen*. So far, we'd only played a couple of opening slots to nobody—our band name certainly was accurate as far as audience numbers went—but I didn't care about that. I was there for the creative camaraderie, and I was having a great time.

Even family relations weren't bad. I'd got back in touch with my father after I started working and could pay back the money he'd loaned me. A truce of sorts had developed with his wife, and she invited me to my little siblings' birthday parties, their school events, and the occasional family dinner. Sometimes, after she'd got a few wines into her, she even apologised for having been such a monster when I was a teenager.

Things were, I thought, going very well.

And then the crying started.

It never happened at work, but at home almost anything set me off. My boyfriend would look at me in horror as I, once again, collapsed into floods of tears that I could neither contain nor explain. He'd say, 'You really need to get some help with this!'

He was right, I did need help, but if anything, his suggestion only made me cry harder. Asking for help was too risky, too shameful, and anyway, I couldn't even admit that I needed it. My child self still implicitly believed that had I ever really needed help, my father would have given it. He hadn't, and therefore I didn't.

*

The crying had been going on for many weeks, but I was still in denial about needing help when something happened that meant I couldn't ignore it any longer. It all started one day when I was taking a break at work and idly looking through some corners of Usenet that I hadn't investigated before. I stumbled across something that would change my life—a group called 'alt.sexual.abuse.recovery'.

Hands shaking, I opened the group. I scanned the post titles and registered the history count, the backlog of posts from hundreds, thousands, who knows how many people. The realisation pounded through my body: *I'm not the only one.*

On a theoretical level, I knew other people were sexually abused, but I'd never talked to them about it, so I didn't know if their experiences were anything like mine. Based on the occasional sensationalised case that made it into the news, I had formed two assumptions. One was that the sexual abuse of children was extremely rare, and the other was that, in all other cases, it was far worse than anything that had happened to me.

The posts on alt.sexual.abuse.recovery showed me the reality. Firstly, there were *so many posts* from people writing about their experiences. Clearly, this crime we had all survived was not as rare as I'd thought.

And secondly, without in any way trying to categorise or compare trauma, posters described a wide range of experiences. I noticed that many prefaced their story with, 'It's not as bad as what others here went through...' I found this diminishment of their horrific experiences both sad and familiar. For so many years, I had been telling myself that I didn't need help because my experiences 'weren't bad enough'. As if any level of child sexual abuse could be considered 'mild', like a common cold, and as easily recovered from.

For the next three days I obsessed over alt.sexual.abuse.recovery. I snatched every moment between urgent work demands, letting the non-urgent jobs slide, to read through the group's backlog. I devoured the stories and the hurts and the hard-won wisdom of total strangers on the net.

And then, on the fourth day, they were all gone.

The group's name was still listed, but when I opened it, it was completely empty. I checked my other regular groups, and they were all fine. I went back into alt.sexual.abuse.recovery... still nothing.

I made my first post there, a sad little, 'Where did everyone go?' My post appeared, all alone in the black void.

Before long, I got a response from a man working at another Australian university. 'AARNET has blocked it,' he told me, referring to the Australian Research Network that was our gateway to the US-based Usenet. 'They've filtered out all groups with the word 'sex' in the title.'

The shock was visceral. Parts of my body dropped through the floor. My hands didn't belong to me. I had forgotten how to breathe.

Somehow, I got myself up and out of the office without drawing attention to myself.

Somehow, my legs walked me to the university's medical clinic.

I made it through the door and then I broke down in front of the receptionist.

'I n-n-need help,' I managed to gasp, 'I've been sexually abused.'

I got lucky. Because this medical practice tended to a student population, there was always an on-duty psychologist, and she could see me straight away. What's more, she turned out to be a specialist on sexual abuse. My rehabilitation started that very day.

*

I cannot even begin to quantify just how much that woman helped me in the year of intensive therapy that followed. She showed me how my child's brain had come up with theories about what was happening to me which, when reassessed from an adult perspective, were understandable but completely wrong. Under her guidance, I learnt how to stop blaming myself for the misfortunes that had befallen me as a child. I learnt how to reach deep inside to comfort that inner child in ways she had so needed but had been denied. And, perhaps most helpful of all, I learnt about cognitive behaviour therapy (CBT). Or, as I liked to think of it, how to troubleshoot my own brain.

My psychologist took the opportunity to teach me CBT the day I came to our session, upset about an interaction at work. It had happened the day before, after Wildthing had suffered an unrecoverable hard drive failure and crashed.

In 1992 we didn't have redundant hard drives and failover servers, so all I could do was open Wildthing up, replace the drive, and restore the data from backup—an unplanned job that took several hours in the middle of a busy working day.

Geoff fielded calls while I worked on the server. Most people understood that it was an unforeseen event and I was doing the best I could—but there was one academic who phoned repeatedly, demanding to speak to me. On his third call, Geoff reluctantly handed over the phone.

'I need the server back immediately!' shouted the academic, 'There are files on there I need for a lecture!'

'I'm working as quickly as I can...' I said.

'You don't understand, you're not LISTENING to me,' he interrupted, 'I need those files NOW!'

Hot rage burst out of the pit of my belly, blazed up through my body and erupted out of my mouth. 'Oh, will you just FUCK OFF,' I snapped and slammed down the phone.

Afterwards, once Wildthing was back up and running, I felt like a piece of shit. I knew the academic had been out of order, but I should not have responded in the way I did. It wasn't professional, it wasn't who I wanted to be. I still felt bad even after I heard the man had complained about me in a faculty meeting, and everyone had laughed at him, and said 'Serves you right for hassling Carol in the middle of an outage!'

After I relayed this story to my therapist, she stood up and wrote on her white board. On one side she wrote 'What he said', and on the other 'How you reacted'. Then she drew a line between them and touched the marker pen near the start of it.

'What were you thinking here?' she asked.

'Oh, um, I don't know... I guess, something like, this is really unfair?' I hazarded.

'This is unfair' she wrote on the board and drew an arrow from it to the point she'd marked.

'And what about here?' she asked, tapping on the line further to the right.

'I don't think there was anything else...' I said.

'Do you normally go directly from thinking something is unfair to outright rage?' she asked, 'No? Well then, there would have been more thoughts. We all have sequences of thoughts in our brain that are so habitual they flash by without us consciously knowing. What we're doing here is trying to pinpoint each thought in the progression, and then we'll come up with ways to retrain them.'

Oh wow, I totally got that! It was bad code! There was a buggy routine embedded in my thought processes, and it had taken me from irritation to explosion in a millisecond.

The therapist continued to press me on what I'd thought next, and over the course of the session we identified the sequence that had led to my malfunction. The final thought bug that had triggered my rage was, 'He despises me'.

'Now we're getting somewhere!' said the therapist, 'Next session, we'll start work on retraining this pattern.'

As I was getting up to leave, she added, 'But Carol, don't be so hard on yourself, it sounds like you're overworked.' I noted what she said, but it didn't really register until much later. Right then, my head was too full of CBT for any more new ideas.

*

CBT isn't easy. Even once you've identified the bad routine, you have to work hard at catching yourself as you slip into its well-worn neural path. Only then can you mindfully direct your thoughts into the healthier routine that you developed earlier. You slip up many times, but if you're lucky enough to have good help, as I was, and if you keep working at it, it does wonders.

Many of my therapy sessions were extremely difficult. We needed to delve into memories that I had never forgotten, but that my mind shied away from facing head on. For a while I was troubled with terrible nightmares and gut-churning flashbacks, but the psychologist gently worked through each one with me, helping to lessen their sting.

Nine or ten months into therapy, when I was feeling far more at peace with myself than I could ever remember, I had a different type of flashback. This time, it was a healing one.

I was at home, sitting on my bed and doing a relaxation exercise that my therapist had taught me. Suddenly, a memory welled up in my mind, so vividly that it took me back into the body of my six-year-old self. It was from a time before my father left, before the bullying, before the abuse and the worst of the neglect.

I was alone, trotting down the rough track in the bushland out the back of our house. As I went, I sang a song that

I had just made up and was very pleased with—it went 'I can be anything in the world, I can be anything in the world'. The memory was exquisitely detailed: the twists and turns of the track through the bush, the sound of my voice as I sang, the warm air as it caressed my skin, and my exact feeling of curious pleasure in the world, and unquestioning confidence about my place in it.

After the flashback subsided, I sat quietly for a while and reflected on this memory. For the first time since that innocent age, I thought, *'It's OK to be me.'*

This felt so significant that right then, I resolved to always remember it. I wanted to never again forget what I'd known when I was six: that there is nothing wrong with me, that I am fine just the way I am.

I had been doing a lot of writing for my therapy homework, all on scraps of used printer paper, scribbled on and thrown away. But this thought, this one called for something I could never discard.

So, I got a tattoo, the only one I have. It's on my left shoulder and it's a picture of the goddess Athena copied from a book I loved when I was that happy six-year-old. I instinctively knew that there would be times in my life when I'd feel unwanted and unworthy and fundamentally ruined, and I hoped that in those moments Athena would remind me that I, like her, am strong and capable and clever.

And twenty-three-year-old Carol was right. There have indeed been many times when a glimpse in the mirror of Athena on my shoulder has brought me out of a self-critical tailspin by reminding me of that powerful shift to self-acceptance. The tattoo didn't fix anything, but it's been a great workaround.

*

And what of alt.sexual.abuse.recovery? It was gone for around six months before the powers-that-be at AARNET relented to community pressure and removed the filter.

In the meantime, I had formed an email support network of two with the man who replied when the group first disappeared. We never met in person, but we told each other our stories and checked in on each other almost every day. I hope our emails were as much help to him as they were to me.

I later heard that the breast cancer support group had also been filtered out in the ban. I understood that the AARNET people thought their filters were fixing the bug of porn infiltrating a research network, but the broad classification of 'sex = porn, breast = porn' suggests a decision made by people with neither breasts, nor a first-hand reason to fear sexual predators.

When your troubleshooting decisions impact so many, it is always worth remembering that one person's bug may be another person's feature.

I'm still tracking down bugs in my computer solutions, and I'm still troubleshooting myself. The therapy I had in my early twenties was life-changing, but the thing about childhood trauma is that it raises different challenges in the different phases of your life. Challenges that you rarely see coming.

I'll be dealing with the consequences of other people's choices all my life, but at least I'm better at asking for help these days. I also have many more tools in my troubleshooting kit—I have loving relationships with my closest family and friends, I have the boundaries I had to set, and I have daily ways to support my nervous system with good nutrition, gentle exercise, sunshine, laughter, and music.

I'm realistic and I know there are bugs that can never be fully resolved. But, whether it's a computer system or myself, it's never a waste of time to hunt down and understand the root causes. And then, even if I can't fully fix the problem, I can at least find some solid workarounds. At the end of the day, that's good troubleshooting.

9. How to be kind to yourself

MONTHS INTO THERAPY, AFTER THE PSYCHOLOGIST had helped me work through the gnarliest of my thought bugs, I returned to the comment she had made after that first CBT lesson:

'But Carol, don't be so hard on yourself, it sounds like you're overworked.'

This had been bothering me. At first, the word 'overwork' had conjured an image of a permanently exhausted minimum-wage battler, working three jobs to barely stay afloat—but that wasn't me. I was one of the lucky ones, an educated person with opportunities and a liveable wage. I had nothing to complain about, right?

However, I had to admit that my to-do list was only getting longer. I was now well-established as the PC go-to person for two hundred staff and post-grads in the School of Electrical Engineering—and while that might not sound like many users, keep in mind that this was a time when all PC repairing, replacing, installing, and configuring had to be done by hand. I was flat out upgrading the fleet to run the new breed of Windows applications, but I was hampered by endless interruptions. My phone rang all day with requests, questions, and cries for help.

When Geoff needed to concentrate, he put his phone on silent and stuck a 'Closed' sign on his desk, but I couldn't bring myself to do that. Two years into the job, I still jumped every time someone called. I felt so grateful to even be there that it seemed the least I could do—but it wasn't just that. The therapist was right, I was being hard on myself. I thought I *should* be able to do it all, that anything less was unacceptable.

And I wasn't the only one. I'd noticed that, just like the academic from the Fuck Off Incident, some of my users were getting very demanding when they called. That wasn't helping my stress levels at all, and it was only a matter of time before I told someone else where to put it.

'It's not surprising,' my therapist said, when I explained all this to her. 'You've *trained* them to expect an immediate response.'

Damn it, I thought, recalling the way I dropped everything whenever someone called, she might just be right.

Possible bug identified, I went straight into trouble-shooting mode. As Geoff had taught me, I changed only one thing, and then I waited to see what happened.

My change was to introduce a delay into my availability. I would no longer answer my phone before 9 am, giving me an hour or so of quiet time to work through my backlog. And when I did answer, unless it was the School Secretary or the Student Admin Officer (two ladies who were even busier than me and whom I always prioritised), I would quickly triage the request, and then I'd put them off, just a bit.

Instead of, 'I'll be right there,' I started saying, 'I can come by at 11:15, how's that?' I wrote the booking in my planner, and because it was my availability I was modifying and not my reliability, I arrived at their desk at 11:15 on the dot.

The results of this experiment astonished me.

Within a couple of weeks, people who had previously been rude and demanding now started their call with, 'Hi, Carol, I know how busy you are, but I'd really appreciate your help...' At the same time, thanks to my early morning quiet time, I was making real progress on my to-do list. And in a win for everyone, I felt much less likely to bite someone's head off.

I found it extraordinary that a simple setting of expectations had so dramatically improved my work life. It was thrilling to realise I wasn't completely powerless, that there were things I could change which, in small but effective ways, improved my situation. Until then, I had worked reactively, my priorities dictated by emergencies and demands, but this had worn me out while making everyone cranky. Now I could see how much better it was to work *pro*actively.

My number one priority became long-term work reduction, even if it meant more work in the short-term. Over the next year, I set up all the staff PCs identically so I wouldn't waste time finding my way around them. I wrote my own tools to automatically push out updates when my users logged in, saving me in-person visits. I put a PC on a trolley, stuck a label on it reading 'Spare Tyre', and wheeled it in whenever I had to remove someone's computer for repairs, buying myself time for the job. And I upgraded Wildthing to a redundant disk controller to protect both of us from another hard drive catastrophe.

One day in my third year on the job, I went to see the Head of School to request funds for a software distribution system that would greatly reduce my users' dependency on me.

I often needed to go cap in hand to the Head—in an environment of continuously tight budgets, he kept a firm grip on all spending. The first time I approached him for money he denied my request, but then took the time to explain how I should ask in the future. It was a great lesson in 'managing up'.

'Your job,' he said, 'is to make my life easier—or at least, not any harder. Don't come here with a whole lot of jargon and expect me to listen. I want a clear explanation of the problem you're trying to solve, and a selection of costed options to choose from. I don't promise you'll always get what you want, but I won't be able to help you if you don't help me.'

He only needed to explain this once—afterwards I always came prepared. I brought examples of the problem and arguments for why it needed to be solved. And I gave him three options: the bargain-basement one, the bells-and-whistles one, and the sensible middle option that I really wanted. I didn't always succeed, but even when he couldn't say yes, the Head always let me know how much he appreciated my efforts.

On the day I asked him to buy the software distribution product, I had not followed my usual pattern. This time, I came with only that option.

The Head sighed and shook his head like a disappointed parent who thought they'd raised you better than this.

'Now Carol, you know I can't justify this expense if you don't give me an idea of the other choices,' he said.

'OK, how's this,' I replied. 'Either we buy this product and everyone can get their applications through self-service, or you hire me a full-time assistant to keep doing the installs by hand.'

I got my software distribution system.

*

Growing up around adults who habitually prioritised their own needs over mine had led me to equate self-care with selfishness—but now I was learning that treating myself with kindness didn't mean neglecting others. Instead, looking out for myself was improving my relationships, and not just at work. There were big changes happening in my personal life too.

Since I'd started dating at sixteen I had been, it's fair to say, terrible at romance. Each relationship, whether a one-night stand or an official girlfriend-boyfriend thing, was chronically unbalanced. Either I was more into him, and I allowed him to mistreat me, or he was more into me, and I despised him for it.

My drinking buddy boyfriend was the former type. I thought he was the coolest guy ever, while he, unemployed and always short of cash, treated me like an ATM with benefits. Thanks to therapy though, I was changing. On my twenty-third birthday I found out he'd been cheating, and I kicked his lying ass to the kerb with the complete conviction that I deserved better. That birthday, the greatest gift I received was the self-respect I gave myself.

Soon after that breakup, I started dating Lobster from the band. My relationship shortcomings hadn't just disappeared overnight, and at first, I was a nightmare. I pulled him in and pushed him away, over and over again, while he waited patiently and without judgement for me to make up my mind. It was a remarkable gift of acceptance, and his patience paid off. Eventually, I gave in to the simple, undramatic pleasures of just being with him.

Lobster was very different to my previous boyfriends. Superficially, they had been university-educated and middle class, while he was a gardener from a proudly working-class family. Far more significantly, Lobster was the first man to love me for exactly who I was.

Our relationship, until it ran its natural course, was delightfully uncomplicated. We were just two people who enjoyed each other's company and had chosen to walk together for a while. And along the way, Lobster modelled something else I'd never experienced: how to care for someone without resentment or an expectation of anything in return. This was by far his greatest gift to me, as it is only through being cared for by another that we learn how to care for ourselves.

It was Lobster who looked after me when I was knocked flat by the double whammy of wisdom teeth removal and an allergic reaction to penicillin. It was Lobster who talked me out of going into work when I wasn't well enough, insisting I give myself time to recover. It was Lobster who got me regularly brushing my teeth, a habit I'd never established in childhood. And it was Lobster who showed me how we could disagree without it turning into a big deal or changing how we felt about each other. By being kind, he taught me how to be kind to myself.

*

The calmer environments I now enjoyed at work and at home allowed something new to happen, something I sensed with a curious wonder. It was an internal coalescence, a coming together of the parts of myself that trauma had split apart.

When I was a young child, after my father left, my sense of self had fractured along with my life. I remember noticing that I felt like a completely different person at his house, at my mother's house, and at school. The Christmas after he left, he returned for the day, and for the first time in a year I was myself again. The bullying and neglect retreated like a nightmare as I laughed and ran around with my father like a normal eight-year-old. It felt light and easy and *right*.

Then he got into his car and drove away, and as I watched him go the heaviness, so briefly lifted, descended again. I knew then that my simple, carefree life was lost forever. Within months it got even worse when the sexual assault on the train created another fracture, later widened by abuse, isolating a part of myself that felt shameful and dirty and ruined.

Now, after the therapy, after I'd wrested my work under control, after I'd learnt how to be loved, I sensed those separate parts knitting themselves back together. I understood then that work and life and love aren't separate; they are interconnected parts of our own unique selves.

I noticed how the same patterns played out in different areas of my life. Just as I had been working reactively, I had also been *living* reactively. I had focussed on financial solvency as a reaction to my years on the edge of poverty. I had sought community amongst the fringe dwellers of the band scene in reaction to a painful history of rejection. I had relationships with men just because they wanted it, my inner outcast fearful of being alone and grateful for the attention, not considering whether he was remotely right for me.

It was clear to me now that the lessons I learnt in one part of my life could be just as applicable in the others.

Lobster insisting, 'Take the day off so you don't make yourself sicker'; the Head instructing, 'Make my life easier—or at least not harder'; me telling my users, 'I'm busy right now but I can come see you in two hours'. They were all the same thing! They were about doing something now that raised my chances of a better future.

I started saying a little mantra to myself whenever I needed a reminder to do the small but effective things today. It is:

'Be kind to your future self.'

It's essentially the same as 'Don't put off until tomorrow what you can do today', but a lot less naggy. A lot kinder.

I'm in my mid-fifties now, and I'm very grateful to my younger self for the kindnesses she showed me. Thanks to her tenacity, I transformed her interest in work efficiency into a specialisation in IT automation. Thanks to her openness to trying something new, I never got stuck in a rut. Thanks to her ability to reflect and grow, I have genuine and unconditional love in my life. And thanks to both her and Lobster, my teeth aren't bad either.

10. How to jump

SOMETIMES, THE KINDEST THING YOU CAN DO FOR yourself is to know when to jump, but that's not all there is to it. When you're contemplating a terrifying leap into the unknown, you must also know *how*.

It was a jump I took, quite literally, while on a hike with my university's Bushwalking Club that gave me a mind hack that proved particularly helpful with a career decision I made at the age of twenty-six.

But before we get to that decision, a quick diversion back to the key moment on that hike.

SATURDAY 3RD OCTOBER 1987, BLUE MOUNTAINS NATIONAL PARK

I stand on the edge of the precipice, staring down at water which looks much further away than seemed possible before I climbed up here. I'd felt so confident way back then, a couple of minutes ago. I'd been watching friends leap from this cliff, laughing at the ones who flapped their arms like would-be wings, taking their brief moment of flight before they splashed into the crystal depths of the water hole. It looked like fun and I wanted to try it too.

Now, at the point of taking my plunge, I am frozen. Other jumpers are coming up behind me, and they hassle me to 'Get a move on' or 'Make up your mind'. My fear wants me to turn tail and scramble back down the way I came, but my pride won't let me, so I prevaricate, torn between the unknown ahead and the humiliation behind. I tell my recalcitrant body, 'It's fine, you know it's fine, you've seen other people doing it, it's perfectly safe.' But it remains stubbornly unmovable.

Eventually, after what feels like an hour but is probably no more than a minute, my willpower wins. My knees bend in preparation, my lungs take in a deep breath, and I leap!

A suspended moment as I hang in the air. The pull of gravity asserting itself. The splash as my body knifes into the cool water. My torpedo momentum slowing as I descend to a gentle touchdown on the muddy floor of the water hole. Then kicking off towards the dappled sunlight above, emerging into the air, my body coursing with adrenaline, laughing with helpless joy as I swim to shore.

And then clambering up the cliff to do it again.

*

I still recall that water jump in a certain type of situation: it's when I have made a perfectly reasonable decision but now, at the point of action, I am terrified.

In those moments, I remember what it was like to stand on that precipice. I remember the exact flavour of my fear, and then I remind myself, just as I did back then, to trust the rational decision I'd made on solid ground. I recall exactly how it felt to push my foot forward through the reluctance, I take a deep breath and then, metaphorically, I jump.

I did a lot of scary things in the years following that eighteen-year-old's first water jump: I found ways to support myself through the rest of my degree, I took chances, I went against the grain. So often did I deliberately recall that jump that I even coined a name for it: I called it my Water Jump Theory of Life.

My WJToL comes from a belief that it's better to do something than nothing. My nervous system wants me to stay safe, to remain within known parameters, to do nothing out of the ordinary—but if I give into it all the time, I risk a lot more. I risk missing unforgettable experiences, limiting my horizons and forgoing life-changing adventures. I miss really *living*.

My WJToL is not about being reckless. At its core is trust in the *rational* and *informed* decision I had made earlier. If new information presents itself at the point of action, I can still return to solid ground and reconsider. However, if the only impediment is my fear, I have a way to push through it.

*

In 1996, my WJToL helped me through the scariest leap of my working life so far. I'd been with the School of Electrical

Engineering for almost five years when I made the decision. Many said I was mad.

It all started the year before, in 1995. With the hectic build phase of the School's staff PC network behind me, and my proactive establishment of good practices and processes, things were ticking along nicely. I'd even been able to take some time off, cutting down to four days a week for a couple of months while the band recorded an album.

The downside was that my work was becoming routine. I knew my job inside out: how to keep my servers humming along, how to rebuild a PC from the motherboard up, how to introduce new applications, and how to monitor and support existing ones.

I also knew all the people in the School who, in some ways, felt more like family than my own ever had. Colleagues invited me into their homes, told me their stories, and offered me advice. And like wise elders they indulged my delayed adolescence, refraining from comment when I turned up to work hungover, tattooed, and with my hair dyed some unnatural colour. They kept on believing in me.

But I had been in this job for more than four years, and that feels like a very long time in your early twenties. I was feeling an itch, I wanted a change.

My first thought was further study, and I applied to do a Master of Information Technology Management with the university's School of Computer Science—but they rejected my application stating that, at twenty-five years old, I couldn't possibly have the requisite five years of relevant work experience. Despite most of it being at this very university.

I was still pondering whether I could be bothered fighting this decision when I was side-tracked by an unexpected phone call from the Executive Assistant to the Dean of the Faculty of Engineering, inviting me to a meeting with him.

I accepted, and then I turned to Geoff.

'That was weird,' I said, 'the Dean wants me to come to a meeting.'

'Uh-huh,' said Geoff, noncommittally. It was only much later I found out that he'd been asked first, and the bastard had dropped me in it instead.

When I arrived for my meeting with the Dean, I found I wasn't the only one to have been summoned. Also present were the IT system administrators from the Schools of Mechanical Engineering and Civil Engineering. They were much older than me, late forties or early fifties, and this was the first time I'd met either of them.

The Dean got quickly down to business.

'We're spending more and more on staff computing,' he said, 'but with each School managing their own computer networks, it's very inefficient. I want us to move towards combining our IT systems across the Faculty—and before you say anything, I know that can't happen overnight. What I want is for the three of you to put your heads together and come up with ways to align what you're doing, with a view to combining services as soon as it's practical.'

He gave us three months to produce a paper with recommendations and a roadmap towards consolidation.

I left the meeting feeling invigorated—I had been looking for a new challenge and maybe this was it. I was looking forward to finding out how Civil and Mechanical, as I'm going to refer to the other sys admins, had set up their computer networks. Perhaps I'd learn some good tricks from my new colleagues.

As it turned out, all they taught me was that older does not necessarily mean wiser.

Civil was a nice enough chap, but very ineffectual. He sat back in our meetings, a comfortable smile on his face, contributing almost nothing.

Mechanical, on the other hand, soon revealed himself to be a pontificating old mansplainer. He made a point of disputing every single thing I said, no matter how bland and uncontroversial it was. I was used to freely exchanging ideas with Geoff, and to the broad respect I enjoyed across the School of Electrical Engineering, so I was perplexed by Mechanical's adversarial attitude.

Looking back, I suspect he felt threatened by his own looming irrelevance, and saw me as a symbol of a computer world that had moved on too rapidly from the mainframe era he'd started in. But I was incapable of such insights back then, and I just knew we had an important job to do. I couldn't understand why both of them, one passively and the other obstructively, were making it so hard.

I pushed against their resistance, asking lots of questions about their computer networks, hoping to find common ground that we could build on. It became apparent that they both presided over IT environments that were haphazardly strung together, hopelessly outdated, and plagued with outages. But Civil just shrugged in a way that said, 'That's how it is, what else can you do?' And Mechanical made it abundantly clear he had nothing to learn from a kid like me.

I ploughed on, determined to achieve the brief assigned to us by the Dean, but when it came to co-authoring the roadmap, I couldn't get either of them to write a damn word. They came empty-handed to each meeting, bringing excuses in place of the sections they'd promised to write. After a few weeks of this, keen to move things along, I offered to author the entire first draft.

For the next week, I laboured over my document, working on it every evening and through the weekend. By the following meeting, I had mapped out all the subject headings and filled in as much detail as I could. It was a long way off finished, but at the very least I figured it would give us a framework for constructive discussions.

How wrong I was: all I achieved with my hard work was to provide Mechanical with ammunition. At our next meetings, he focused on trivialities in my document while refusing to discuss either the big picture or the technical detail.

'I don't agree with your use of the word "while" in the introductory paragraph,' he droned. 'No, no, this *is* important,' he continued over my protests, 'and we *will* discuss it now. I am also not happy about how you've used the word "and" in the following sentence.' All the while, Civil sat back with that placid smile on his face, contributing nothing.

I was starting to think that neither of them had a bloody clue.

I complained to Geoff, but he just laughed and said, 'And now you know why I refused when the Dean asked me!' (See, told you he was a bastard.)

I was now dreading the meetings, all my earlier enthusiasm knocked out of me—but I was such an old hand, going back to primary school, at showing up in places where I knew I'd be utterly miserable, that it didn't even occur to me to quit the project.

Eventually, the tortuous three months came to an end. We had a document, but only because I'd written the whole thing myself. There must have been a meeting where we presented it to the Dean, but I've blocked it out. All I wanted was to put the whole frustrating affair behind me, to go back to the collegiate environment of the School of Electrical Engineering and leave Civil and Mechanical to their own stupid devices.

But the Dean wasn't done with me yet. Again, I was summoned to his office, and this time it was only me.

'We have a few problems with our computers here in the Faculty Office,' he told me. 'One of the printers hasn't worked for months, and we're always having trouble with email. Could you take a look?'

I was happy to, and within an hour, I'd fixed the printer and several other issues. I was about to leave when the Dean called me back.

'Civil Engineering has been looking after our IT for the last few years, but I'd like you to take it over,' he said. 'I've discussed it with your Head of School and he's agreed, as long as you're happy to take on the extra workload?'

So, this support request had been a test. Even so, the Dean seemed like a nice chap and it was only another five users, so I said, 'Sure, no problem.'

Over the next few weeks, I upgraded their PCs to my standard and hooked them up to my network services. While I was working in the Faculty Office one day, the Dean again called me over.

'I was wondering if you have put much thought into your future. Do you know what you want to do in your longer-term career?' he asked.

'I'm not sure,' I said. 'I applied to do a Masters of IT Management with the School of Computer Science, but they knocked me back. I haven't thought any further about it.' And then I went back to my work without wondering why he'd asked.

The answer came a couple of weeks later when, instead of a request, the Dean had an offer for me.

'I am creating a new position of IT Manager for the whole Faculty of Engineering,' he said, 'and I'm offering it to you.'

It all makes sense in retrospect. The Dean knew that the alignment of IT services across the Faculty would never happen organically, and that it would take top-down authority to override Civil and Mechanical and get anything done. He would have figured out who had written the proposal, and now he also had first-hand experience of the IT service my users enjoyed. But I hadn't joined those dots back then, and I was completely stunned by his offer.

The Dean went on, outlining the benefits. There would be a substantial pay rise, and that was just the beginning. 'I have spoken to the Dean of Computer Science, and you have now been accepted into the master's program. We will cover your fees, and you will have paid time off to study.'

It was an unbelievably generous package, and I felt honoured and grateful... but something stopped me from accepting right away.

'Can I have a week to think about it?' I asked.

A thought had risen unbidden in my mind, and as I walked away from the Dean's office, I couldn't shake it. 'If I stay here,' I thought, 'I'll become institutionalised and resistant to change. Just like Civil and Mechanical.'

For the next week I mulled over the offer. The pros were clear: support through a master's degree, more money, remaining in a familiar place with people I liked. And my sense of loyalty was putting up a strong argument too. The School of Electrical Engineering had taken a chance on me when I was little more than a mixed-up kid. With their support I had grown, both professionally and personally, into the type of functioning adult who could be offered such a big promotion.

I had only one item on the con list: a vision of myself at fifty years old, still here, bitter and frustrated, trying to prove I was worth my tenured position by obstructing some earnest young person who was just trying to do a good job. What a nightmare.

I realised something then, and once I'd thought it, there was no going back. As much as it scared me, I was going to have to leave the nest if I wanted to fly.

*

When the week was up, I booked two meetings. The first was with the Dean. I thanked him for his kind offer, and then I turned him down.

The second was with my Head of School. This one was much harder as I admired him a great deal. I appreciated the time he had taken to encourage and educate me, and how he always treated me as a valued colleague and not a subordinate. I really didn't want to disappoint him and my reluctance intensified as I walked towards his office. So, I deliberately called on my Water Jump Theory of Life, summoning that clifftop memory to push one foot in front of the other until I arrived at his door.

The Head waved me into his office, and I sat down in my familiar place across from him at his great big desk.

He regarded me steadily as I took a deep breath in preparation for the jump.

'I'm resigning,' I said, 'It won't be straight away, I'll give you time to find a replacement for me, but it's time for me to go.'

The Head sighed and nodded, like he'd been expecting this. He didn't try to change my mind, but he did comment, 'There's a lot more to life than money you know.'

'That's not why I'm leaving,' I said, 'it's just, I have to see what's out there.'

I set my leaving date as 1st April 1996. It felt like a nice bookend, exactly five years after that April Fool's Day when I'd started.

In my final months, I finished up a few projects, sat in on interviews for my replacement, and trained the young chap they hired. I did not search for a new job. Perhaps my Water Jump thinking had influenced me too much, because now that I'd pushed through the fear of leaving my comfort zone, I found I didn't need to know where I would land. I just wanted to leap out into the air and see what happened.

11. How to be in it for the money

WHEN THE HEAD SAID THAT THERE IS MORE TO life than money, I thoroughly agreed with the sentiment, but I would have added, 'when you have enough'. I could well remember what it was like to not have enough money, but now, after five years of stable work, I had a comfortable buffer of savings in the bank.

Working at a public university, my pay had been lower than average for an IT professional, but in my personal life I was surrounded by friends who earnt much less than me, so I felt wealthy. Like them, I lived cheaply: cooking vegetarian meals at home, buying second hand where possible, low-cost nights out at my local pub seeing bands and drinking $3 beers. I never felt I lacked for anything.

After I quit my job, I treated myself to a nice long break. I made very few plans. I wanted to wake each day without an agenda and then decide what I felt like doing.

It took a couple of weeks to get used to my new unscheduled life, and then it was bliss. Lobster and I went to see bands almost every night, and on weekends we drove up the coast and camped in sand dunes behind deserted beaches. I caught up with friends, and I played my violin, and as the tension dissipated from my body I discovered, to my surprise, that my shoulders weren't naturally rock hard after all.

The last time I'd had such a long period without work was the five months I'd been unemployed after my Honours year. My memories of that time were dark: the long, drug-addled nights, the constant worries, the persistent underlying sadness. This break, in contrast, was filled with sunshine and music and joy. This time, I could enjoy the more there is to life, because I had enough.

However the bank balance would only stretch so far, and after three months it was time to get back to work. In July 1996, I got my hair cut, bought a suit, updated my CV, and started job hunting. I wanted to find work that extended my technical skills, and I was keen to explore sectors outside education.

Within a couple of weeks I was offered a contract at, let's call it, 'Money Bank'. The hourly rate was, for me, eye-wateringly huge. I knew that universities paid below market rate, but this was literally double what I'd been taking home before! Having never had the means to develop spendthrift habits, my only plan for this windfall was to enjoy watching it stack up in my bank account.

The job was to be a technical resource on a Windows 95 upgrade project, something I already had experience with. 'It was the main reason we hired you,' the Project Manager said after I started at Money Bank, 'you're the only one here who's already done a Windows 95 roll-out.'

'Cool,' I thought to myself, 'I'll be able to give them some good value for all this crazy money they're paying me.'

The team consisted of the Project Manager, a Tech Lead, and eight of us contractors, hired to do the grunt work of upgrading thousands of office PCs. We sat together in the Project Office which had been set up, very oddly, like a classroom. The PM and the Tech Lead sat at the front with their backs to the only window, while all of us contractors sat facing them, going back in rows of two. I was the last hire and took the only remaining free desk. I was also the only woman, but by then, that felt so normal that I barely noticed.

The PM gave me the Project Plan from which I learned we would be deploying a custom build of Windows 95, pre-loaded with every application dependency and printer driver that anyone in the Bank might need. This was, according to the Plan's effusive intro, *the* solution to the problem of escalating IT support costs.

In the mid-nineties, installing an application on a PC was a manual and often fiddly exercise. There would be a list of dependencies you had to install first, with care needed to get the right versions and configuration settings. Printing was tricky too, with a specific driver for each type of printer, and additional steps to get applications to recognise them. Installing a new application or printer could be complex and time-consuming—frustrating for both the IT technician and the end user.

Money Bank sought to address this by packing every possible printer driver and application dependency into a custom build of Windows 95. So, instead of installing vanilla Windows, followed by repeat visits to get applications and printing working, the proposed build would have everything pre-installed. New applications would work effortlessly, all network printers would be accessible, users would be happier, spiralling IT support costs would be reined in, and a new utopia of smoothly efficient IT operations would be ushered in.

It wasn't going well.

When I joined the team, the project was at a standstill. A month or so earlier the super build had been deployed in one small business area, the pilot group, and that's as far as they'd got. We were supposed to be cracking through upgrading the thousands of PCs in the rest of the bank, but there was an issue, and it was preventing us from moving forward to the next group.

Essentially, an application used in the pilot group needed something that was total poison to an application used in the next group. The Tech Lead had been stuck for weeks trying to get these dependencies to coexist—and that was for just two applications. There were hundreds used in Money Bank, both off-the-shelf and custom-made, and only a fraction of them had been worked into the build so far.

'Let me get this straight,' I said to the Tech Lead, 'You're trying to make it so the build can run both of these applications, but they're used by different sections, and *no one actually runs them both anyway?*'

'That's right,' he said.

'Why don't you just have a different build for each section?' I asked, entirely reasonably I thought.

'NO,' he said, 'that's NOT what we promised. The build will be able to run any application used in the bank!'

It all seemed very impractical, but I was open to being proved wrong, so I asked if I could help with troubleshooting bugs or looking for workarounds. He brushed me off. He had complete control of the build, and it looked like he planned to keep it that way, even while eight expensive contractors sat around with nothing to do.

The Project Manager could have intervened, but he always deferred to the Tech Lead. The PM was young, not much older than me, and this was his first IT project. He was completely out of his depth, but he looked good in his fancy suit and expensive watch while floundering, so I expect he went on to have a stellar corporate career.

All the guys in that room looked much more the part than I did. They were decked out in square-shouldered suits, tastefully colourful ties, designer leather shoes and gel-sculpted hair. Meanwhile, I wore a black trouser suit in some sort of stretchy fabric, a shirt in a neutral colour, and my comfortable Blundstone boots. My hair, as always, was how it had landed after a post-shower run-through with a brush, and I never wore makeup.

What I had failed to understand in that money-centric business was that we were all supposed to look rich. Silly me, I just thought I was there to do a job.

And it was quickly driving me nuts that there was no job for me to do. We sat at our computers eight hours a day, trying to look busy, but the only person with actual work was the Tech Lead. One contractor was openly running his own business, using his desk phone to call his customers. Apparently, that was fine. But if I suggested to the PM that I might as well leave early and not bill for a full day, he'd say, 'What's this? *Public service hours?*' His snobbery rankled.

I had been flat out making the absolute most from limited resources when I worked at the university. All I saw here was arrogance and waste.

It was at Money Bank I learned that if I am not given enough to do, I will go looking, and I'll get myself into trouble in the process. One day, I passed the time searching for security holes in the bank's NetWare servers. I didn't have the administrator access I was used to, but even so I found vulnerabilities.

I took a list to the NetWare administrator, naively thinking he'd appreciate my helpfulness.

'What do you expect me to do with this?' he snapped, barely glancing at it.

'I just thought, you see, I've done a lot of NetWare administration, and there's a couple of well-known security holes you've left open here, it would be pretty easy for someone to get in...'

'You don't get it,' he said, 'People around here, they don't understand anything about IT. They're never happy, they always want more and I can never set things up right. If they just got off the servers then everything would be fine!'

'Wait a sec,' I asked, surprised by the direction he'd gone, 'what's more important, users or servers?'

'Why servers of course!' he blustered.

For a second I was speechless. 'Umm, you know, without the users there wouldn't be any servers,' I pointed out. But he'd had quite enough of me by then, and he told me to get lost.

Chastised, I returned to my desk and tried to find a way to look busy that didn't piss anyone off.

A few days later, a service request came in from the pilot group. Probably hoping to shut me up by giving me something to do, the PM sent me to attend to it. After resolving the issue, the section manager wanted to talk.

'Thanks for fixing that, but you know I'm really not happy with the way this project is going,' she told me, 'I don't see how this is going to help my business.'

I wasn't at all sure what she meant, but I tried to placate her by saying, 'I think the idea is to reduce costs across the whole Bank...'

'Perhaps you don't understand,' she said, 'that Money Bank is made up of sections that are run as independent businesses. We each have our own profit margins, and we don't like paying for costs we can't control. This project is going to benefit the larger sections much more than mine and it's not fair that I'm locked into paying for it when I could be getting my IT services far more cheaply on the open market.'

Her comment gave me an idea of the conflicting pressures on this project. While escalating demands were exasperating IT staff to the point of throwing up their hands like that NetWare administrator, the increasing costs of meeting those demands were infuriating the business.

The Windows 95 project had promised a solution that would be both simpler and cheaper than anything the bank currently had, but this promise had been a huge gamble. While the concept of a 'standard build' wasn't new—I had developed them myself in my previous job—no one in the world had ever created the exact build with the exact combination of application dependencies that they were trying to do. This highly risky venture had not only the reputation, but quite possibly the future of the in-house IT Department at stake.

I, once again, tried to be helpful.

'At my old job,' I told the PM and the Tech Lead, 'I set up a system where users could see all the application icons, and if they clicked one they didn't already have, a script ran that installed it from the network drive. I ran it like that with Windows 3 for years and it worked great. I had to adapt it for Windows 95, but I got it working in much the same way. I can show you what I did if you like.'

They heard me out politely, and then the PM said, 'I'm sure that was fine for a *university*, but it's just not *Money Bank*.'

I was becoming accustomed to hearing *Money Bank* used as an adjective, with anything and anyone judged that way. To be full-time employed, as opposed to a short-term contractor like me, you underwent several tests that included a personality profile. I'd hear people say, 'He seemed such a good candidate, but he failed the personality test. He just wasn't *Money Bank*.' I'd arc up, interjecting, 'How in hell do you *fail* a personality test?' They'd give me a look which said, 'We know *you'd* fail.' Bloody would have too.

*

After a couple of frustratingly idle months the PM, perhaps to stop me asking more awkward questions, moved me to Desktop Support. I still officially worked for the project and was only supposed to respond to problems reported by Windows 95 users. A handful more had been installed, but there weren't enough to keep me anywhere near busy. Meanwhile, I could see how flat-out everyone in Desktop Support was. I couldn't just sit around and twiddle my thumbs when there was work to be done, so I started taking tickets off the general queue. The manager of Desktop Support was more than happy for me to do this. At least, she was until the day I took a particular ticket.

This call brought me to one of those offices where an assistant sits guard out the front. 'She can't print,' the assistant said, and waved me straight through.

As I entered, a woman stood up from her desk. She was middle-aged, slim, dressed in a tan pencil skirt and a cream silk shirt. She had immaculate makeup and blow waved, elegantly highlighted hair. I was dressed, as usual, in my bargain basement facsimile of corporate attire, my face bare and my hair air-dried.

I had, by then, experimented with a salon wash-and-blow-dry to get my hair looking more like the women I saw around the Money Bank offices—but when the stylist finished with me, all I saw in the mirror was a total fraud. I washed out the blow wave the moment I got home and there was never any danger of me wasting that sort of time and money again. Anyway, it wasn't like men spent every other morning in a hair salon. Looking rich enough for Money Bank was expensive, and especially so for the women who worked there. The bank may have paid well, but I wondered what percentage of their income went on maintaining that image of perfection.

This woman was older than most I had seen around the bank. There was a steely edge to her deliberate elegance, giving an impression of someone in absolute control of her body, her face, and every immobilised hair on her head. A woman not to be messed with.

She waved her hand at a printer sitting on a low cupboard. 'It doesn't work.' she said. 'Nothing comes out.'

I checked it had paper and toner and was correctly plugged in, while she returned to her PC. Finding nothing wrong I said, 'Excuse me, I'm going to have to check your computer, perhaps I'll need to update the printer driver or something.'

'Perhaps!!' she exploded, 'What do you mean PERHAPS?? Don't you KNOW?!'

I was momentarily stunned by this outburst, but then hot blood flushed my face. 'No,' I said between gritted teeth, 'I do not *know*, that's why I have to look on your computer to *find out*.'

'This is just not GOOD ENOUGH!' she shrieked.

'Well fix it yourself then,' I retorted, and stomped out.

By the time I returned to my desk, she had already rung the Service Desk to register her outrage at the appalling way she'd been treated. The manager of Desktop Support scolded me, 'You can't speak to people like that!' When I protested that the woman was rude to me first, the manager said, 'Yes, yes, I know, she's often rude, she's had people around here in tears before, but that's no excuse. She's a high-level executive and we just have to take it.'

The manager dispatched Jason, a good looking and nicely dressed young man with a playfully camp demeanour, to see to the printing issue. When Jason returned smiling and unscathed, I asked him what his secret was.

'I just tell them how much I love their shoes and notice when they've changed their hair,' he explained, 'and they love me.'

Nice that it works for him, I thought, but this was clearly not a technique I could get away with. One look at my Blundstone boots and any shoe comments from me would be obviously egregious.

*

I was never going to fit in at Money Bank, and it wasn't just my lack of coiffed hair and spikey heels. For starters, they were all *obsessed* with money. The sentiment 'There's more to life than money' would not have resonated with this lot.

Everyone wanted a house near the beach or the latest model BMW, and they deeply resented anything that delayed them getting it. 'Do you KNOW how much tax I pay?' one woman thundered at me, after I'd questioned her statement that all education should be privatised. Once again, I was rendered speechless by a viewpoint I'd never encountered before. Surely, I thought, if she was paying a lot of tax then that meant she earnt a lot. Why couldn't she see how privileged that made her? Why wasn't it enough?

Also, I could not get on board with the presenteeism. I filled in timesheets and billed by the hour, so it wasn't like I was robbing anyone when I left at 5pm—but there was always a sideways glance or a disapproving head shake as I was, once again, the first to head for the door.

The manager of Desktop Support wouldn't even take time off when she was sick. She caught a nasty virus and proceeded to drag herself in day after day, dosed up on cold-and-flu tablets that did nothing for her red eyes, streaming nose, and increasingly croaky voice. 'Go home,' I said to her, more than once. She wouldn't hear of it.

After a week or so, she became so sick that she had no choice but to take a day off—but the very next she was back, still looking like death. She repeated this pattern until she became so ill that her husband had to phone in to say she had a severe chest infection. If I'd needed any proof of Lobster's assertion that 'you'll make it worse if you don't give yourself time to recover,' here it was.

I'd seen enough by then to blame the workplace culture for the way the manager over-exerted herself. Being a *Money Bank* person came with intense pressure to present, at all times, as a type-A go-getter who made work their number-one priority and had absolutely no 'weaknesses'.

'This place is toxic,' I thought. 'It is literally making people sick.'

I can't remember what the last straw was—I'd probably been told off again for saying the wrong thing or looking the wrong way or for having worked at a university like a *public servant*—but I do remember staring at my desk and thinking, 'What the fuck am I doing here? No amount of money is worth this bullshit.'

I needed no Water Jump resolve for my next action—instead it felt like the only thing I could do. I picked up my stuff, walked out, and rang my contracting agent from a pay phone. 'I'm not going back,' I told him.

My time at Money Bank felt like a complete waste. Yes, I'd been paid a lot, but I'd just spent five months being bored, frustrated and disparaged, and I hadn't even picked up a single new technical skill.

I had, however, learnt one thing about myself, and it was something I was quite proud of. In walking away from that outsized pay cheque, I had proved I didn't have a price. I knew now that I was not prepared to sacrifice the fundamentals of who I was for more money than I needed. Once I had enough to live on, my priorities were interesting work, a supportive and accepting environment, and enough time for the rest of my life.

And I reckoned that made my chances of a happy and healthy work life far higher than if I'd only been in it for the money.

12. How to be the business

IT DIDN'T TAKE ME LONG TO LINE UP INTERVIEWS after I walked out of Money Bank. It was now 1997, and in the fifteen years since IBM-compatible personal computers had entered the market, the PC had become *the* essential tool for all white-collar work. As a result, people like me were in high demand.

I had a couple of interviews which didn't go anywhere. In one, the guy asked pop quiz questions, like 'What's the bitmask of a Class C network?' I explained that I didn't know because at the university, we'd had a Class A network and used an unusual bitmask. I knew exactly what the bitmask was *for*, but he didn't want to hear about that. He shook his head dismissively while making a mark in his notebook and there was no offer from him. I would have turned him down anyway. Someone who thought my technical ability was measured in recall of random facts instead of the way I could figure things out was not someone I wanted to work for.

The next interview was with Lew, a director at a company called Australian Communications Management (AusCM), and it felt better from the start. He began by asking what types of work I most enjoyed doing, which I found infinitely more engaging than a quiz. I talked with unfeigned enthusiasm about my time in the School of Electrical Engineering, how I'd built systems and found creative ways to support my users with the limited resources available to me.

Lew then told me a bit about AusCM. It had begun as a branch of a large construction company, set up to do the IT parts of corporate building projects—mostly network cabling and computer relocations. After a few years, with the blessing of the parent company, the branch had broken away to form their own firm.

'We started out doing the same sort of things we'd done on the building projects,' explained Lew, 'but we found that customers want a lot more than cabling. They want servers and software solutions and training, and they want to get it all in one place. We're well covered for network technicians, but now I'm looking for people who can design and build servers. Does that sound like something you'd be interested in?'

What it sounded like was getting away from angry users while I did the fun work behind the scenes. I'd get to learn new tech and build systems and then hand them over to someone else to run while I moved on to the next project.

'It sounds great!' I said.

I was feeling confident I had this all wrapped up when the difficult question came.

'I notice that you didn't put down a referee from Money Bank,' said Lew, 'Why is that?'

Damn.

'Umm, yeah, I know, umm, you see, I moved around a bit there, and I didn't really get to know anyone well enough...'

'I know people at Money Bank,' Lew said, 'and I could ask about you if I wanted to.' He gave me a long look. 'But I won't do that.'

I exhaled, not even aware I'd been holding my breath.

The moment passed. Lew went on to talk about pay and conditions and ask when I could start. It was less pay than I'd been getting at Money Bank, but as I'd proved to myself, that wasn't my priority. I liked Lew, I liked the idea of project work, and my gut was telling me this was the right job. I accepted his offer on the spot.

As I was about to leave, Lew said, 'Just one more thing, I promise you'll never be bored here.'

His parting remark stayed with me, and I mulled over it all evening. I couldn't remember saying anything about boredom in the interview, so it seemed odd. Had he, I wondered, already rung one of his contacts at Money Bank? Had they told him that I had a terrible attitude and even worse footwear, but perhaps also admitted that I'd been frustrated by the lack of interesting work? And had Lew seen that my need to be busy and useful made me a perfect fit for the high-pressure business of commercial project delivery? If so, he was right.

*

When I started work a week later, Lew assigned me to a project building a Lotus Notes network for a large and geographically dispersed organisation.

'I don't have any experience with Notes,' I admitted, worried that I'd be out of the job before I'd even begun.

'Yes, I know,' said Lew, 'that's why I've booked you on the training courses. They start in a couple of weeks. You'll need to pass some exams as well—the customers like seeing that our people have the right certifications.'

Wow, actual training! Back at the School of Electrical Engineering, I had passed seven exams to become a Certified NetWare Engineer, but I'd had to study for them on my own as there was no money for formal instruction. Now I was being offered training as soon as I walked through the door!

This was my first job in commercial IT, as opposed to internal IT, and I sensed that something fundamental had shifted. In my past jobs, I was a cost. I didn't contribute directly to the revenue-raising side of the business, and instead I was an expense of doing that business. At Money Bank, a highly begrudged expense.

Now I *was* the business. Without consultants like me, AusCM had nothing to sell. In booking my training Lew wasn't doing me a favour, he was investing in his product. I decided on the spot that I was going to like being the business.

'Also,' added Lew, 'we didn't have any Notes people when we bid for this project, so we partnered with another company for the design. I want you to shadow their Notes expert and learn as much from him as possible.'

Even better, not just classroom but on-the-job training too! Only ten minutes in and AusCM was already looking a hundred times better than Money Bank.

Next, I met with the Project Manager. I knew nothing about commercial projects, but if I was going to be the business, I had to understand how the business worked. The PM started by walking me through the project plan.

He unrolled an enormous chart, printed across several A1 sheets of paper and taped together. It was a Gantt chart, the first I had ever seen. Down the left side were listed all the project tasks, and across the top were dates. Coloured lines indicated the start and end of each task, with connectors linking dependencies, so that the lines stepped rightwards and down across the page.

I gaped speechlessly at the scale and detail of his chart. I'd done some projects of my own, but I had just set myself a goal and then figured things out as I went. In contrast, this was so *organised!* So *professional!* It made my as-you-go approach look like child's play.

It was only later, when I worked on bids myself, that I realised just how much of these charts are total guesses. Maybe we'd get the effort estimates right if we did the exact same project with the exact same team five times in a row—but the fact is, in IT you never do the same project twice. I didn't know that yet, so I was deeply impressed.

The PM pointed out the boundaries that split the schedule into distinct phases, each ending with a payment milestone.

'The milestones are the really important things,' he said. 'We're not going to get very far if we don't get paid.'

The project was about to start with a short Kick-off phase where we'd do introductions, sort out access, and finalise paperwork. This was followed by Conceptual Design, where we'd agree on high level functionality and project schedule. Next, we'd go into Detailed Design, followed by Build and Test. Then came a particularly critical milestone: the customer had to accept the solution before we could move on to Implementation.

'It's particularly important we don't get stuck here,' said the PM, tapping this milestone symbol. 'I've seen it happen—the customer won't accept the solution because of something or other, and we go forward and back on it for weeks and completely mess up our implementation plans.'

I now know to call what I was looking at a 'Waterfall' schedule, named for the way the phases, and their internal steps, cascade from one to the next. In a time before standardised IT project methodologies, AusCM had developed their own, taking inspiration from the building industry that had birthed them. You could draw direct parallels to the phases of building a house, from conception through to getting your keys.

These days I understand the fundamental weakness of Waterfall: it assumes every important detail can be itemised and planned in advance. Perhaps that's true when building an ordinary house, but let me tell you, from my vantage point three decades later, it's never true for IT projects.

On that first day at AusCM, I took the chart to be as good as gospel. Likewise, I had no qualms when the PM said this complex, year-long project had been sold at a fixed price, set months earlier in a competitive Request for Quotation (RFQ) bidding war. I accepted that this was how business was done.

I was about to learn what a difficult business I had entered, starting with the fundamental reality of fixed price that customers will choose the cheapest bid, but expect the premium offering.

*

'We said we wanted Solaris servers, and that means Sun hardware!' insisted the customer in our second project meeting.

'Your RFQ didn't specify Sun, only the Solaris operating system. These clone servers run Solaris just fine, so I don't see the problem,' parried the PM.

I, the junior in the room, watched silently while this continued, both of them refusing to budge. In the taxi after the meeting, I asked the PM what was going on.

'We only won this project because we put in the lowest bid, and we did that in part by costing a clone brand instead of Sun,' he explained.

The customer came to the next meeting prepared for battle, and he challenged every one of the clone brand's requirements for warranty and support. The PM dialled in the clone salesman who made sweeping promises to meet anything that the much larger Sun Microsystems could offer.

I thought that would settle the matter, but the customer wasn't done and the Server Battle raged on for weeks. I would have missed most of it as I was off doing my Notes training, but somehow the clone salesman got hold of my work mobile number (the first mobile phone I ever had) and rang me almost daily.

'They're great servers, really good quality, and we can do everything they're asking for!' he'd tell me yet again. And I'd reply, yet again, that I had no say in the matter. I was a rookie in this business, but even I could see he must have been desperate if he was wasting his time on me.

Eventually, to my relief, his calls stopped. After I returned to the project, brand new Certified Lotus Professional certificate in hand, I learnt he'd won the Server Battle. With the argument now settled in favour of the clone, we could move on with a schedule that was already running five weeks behind.

While I hadn't yet done much on the project, it had already been a slog for the PM. His strained demeanour gave me another clue about commercial IT: the technical work may be the meat of the project, but it's sandwiched between a lot of non-technical effort. He'd been wrangling, compromising, rescheduling, and recosting, and it looked like hard work of the sort I never wanted to do. I was relieved we were finally getting to the Design Phase where we'd have something far simpler to talk about: the tech.

*

My job was now to shadow the Notes expert from the partner company, learning as much from him as I could. Mr Notes, as I'll call him, was a tall, hefty guy in his early thirties with plenty of opinions and little filter. 'Netware is dead,' he asserted when I told him some of my background, 'and forget about Microsoft, don't waste your time on them. The future is Lotus Notes!' This seemed unlikely, but he looked like the sort who'd relish an argument, so I kept my thoughts to myself.

A couple of days into the Design Phase I got a call from him.

'I've got a meeting with the customer's network manager,' he said. 'If you want to tag along, I can swing by your office and pick you up in ten.'

'I'm not sure where you can stop outside the building...' I began, thinking of the busy, central Sydney thoroughfare twenty-one floors below the AusCM offices.

'Don't worry about that, just be outside when I get there,' he said, and hung up.

Ten minutes later I was standing on the pavement outside, still wondering where Mr Notes was going to stop, when a car cut in front of a taxi and slammed to a halt right in front of me, causing the taxi to swerve into the next lane, its horn blaring loudly. I could see Mr Notes gesturing to me impatiently from the driver's seat, apparently oblivious to the taxi, and to the horns of the other cars driving up to find him stopped in their lane. I ran to his car and jumped into the passenger seat.

Mr Notes took off while I was still buckling my seatbelt, ran the next traffic light just after it turned red, swerving sideways through the intersection towards the centre lane on the other side. The whole time, between revving hard, under-cutting, and yelling at other road users, he talked.

'Listen, I don't know what game your lot were playing with those clone servers, but anyway, it's sorted out now and we need to finalise the server order today. *HEY! GET A MOVE ON!!* I need to see what their network capacity is between sites because some might be able to share servers. *Oh bloody HELL, you could have gone through that light you MORON!* It's a lot of servers if they get one per site—more than we planned for, so we'll have to work this out. *What do you think this is, a bloody SUNDAY DRIVE!!*'

I said nothing, gripping the sides of my seat as he screeched around corners and roared up behind any driver inconsiderate enough to be in front of him. I was queasy and shaky and glad to be alive when we arrived at our destination.

A couple of hours going through network diagrams and WAN link capacities was enough to soothe my frazzled nerves—in fact, it was so tedious that I would have welcomed a nap. Instead, when we were done, I had to follow Mr Notes back out to the car park.

I'm not much of a car person and I couldn't tell you what type he was driving, but I had noticed his car phone. In 1997, car phones had been around longer than mobile phones but were so expensive that they were more of a status symbol. They signalled importance, someone so critical to the business they had to be contactable even while driving. And Mr Notes was not going to let me miss this.

As he backed out of the parking spot, he was already stabbing at the buttons on the phone, making a call. Within a minute, he was shouting at the disembodied voice on the car phone, while all the while continuing to drive like an absolute menace. Mr Notes was, like me, a technical consultant—he was also *the business*—but it seemed to have gone to his head to a dangerous degree. I was supposed to be learning from him, but so far, he wasn't giving me anything I wished to emulate.

A few days later I quashed my misgivings and accepted one more ride with Mr Notes, but it was exactly the same: he drove with careless aggression, he made phone calls, he shouted. After that, I arranged my own transport. This would later prove to be a wise decision.

*

Not long into the Design Phase, we hit another dispute, this time about the project's email component.

'We have to have position-based mailboxes,' insisted the customer, 'this is an essential requirement.'

'But that's not how it works!' protested Mr Notes. 'You have one mailbox per person, and I really recommend you don't change that!'

The customer was adamant. 'This is non-negotiable, it's how we operate. We don't want to email the *person*; we want to email the *post*. We shouldn't have to know who is currently in a role, and whoever it is, they need to see all the past communications to that post.'

'There was nothing about this in the RFQ...' started the PM, but the customer cut him off.

'The RFQ stated that the new email system had to have at least equivalent functionality to our existing one, and this is how our existing email works. You agreed to that in your response. If you didn't understand how our existing system worked, you should have asked before responding.'

Everyone looked at each other in a silent standoff. The customer, a sternly regimented and hyper self-contained type, allowed himself a hint of a victorious smile.

I'd recently learned that the customer had lost the Server Battle, not because the salesman wore him down, but because our PM had gone back to the AusCM bid response and found this assumption:

'Where the customer has not identified a preferred supplier, the provider will select the supplier for any hardware, software, or other material required to deliver the contracted solution.'

This clause had cornered the customer. Either he accepted AusCM's choice of vendor, or he went to his own management, admitted he'd overlooked something, and begged more money for the name brand.

The customer had taken the hit and backed down, but he was clearly still smarting. Now, his little smile betrayed how much he was enjoying his revenge. It had not occurred to anyone to explicitly write, 'We assume the email system will be set up according to standard practice and there won't be anything weird like position-based mailboxes.' He had us on that one.

Since that project, I have written many lists of assumptions, and let me tell you, it's hard. I do my best to pre-empt surprises like that mailbox one, but you try listing all the stuff you're unconsciously assuming. Like, when you walk through your local park you assume you won't get trapped in a Wuthering Heights flash mob, but what if you do? And what if the Kate Bush cosplayers surround you, force you into a red dress and long wig, and make you wuther along with them until you pay each of them ten dollars?

I know this is a ridiculous analogy, but I've seen this Game of Assumptions get equally ridiculous. Even the catch-all clause that 'Everything not explicitly in scope is out of scope' does not protect you from a loosely worded requirement. The customer hadn't known there were non-Sun, Solaris clones, and we hadn't known that email could be set up in any way other than per-person. We were one-all in the Game of Assumptions, where the loser always pays.

Back in the standoff, Mr Notes was the first to break. Sighing, he said, 'I guess we could do something with shared mailboxes...'

'We'll talk about it later,' the PM cut in. Then, turning to the customer, 'And we'll have an answer for you very soon.'

As we returned to AusCM, the PM shook his head over the extra delays and uncosted work to meet this new complication. Then he shrugged and said, 'You win some, you lose some.'

*

We battled on, the PM looking more strained by the day as he shuffled and re-shuffled the schedule and watched the already-thin profit margin dwindling away.

Meanwhile, I had become a scribe for Mr Notes. Like many of the technical people I have worked with before and since, he was incapable of writing anything beyond bullet points, but that would not satisfy our contractual obligations. Documents were tied to milestone payments, and it was falling to me to write them.

I watched over his shoulder, taking notes while he built a test environment and experimented with configurations. My training proved invaluable as I could ask him questions using the right jargon—a generally reliable way to tame an impatient techie.

Fortunately, Mr Notes was a lot less impatient when sitting at a keyboard than he was at a steering wheel. I disliked his arrogance and I hated his road rage, but I respected his technical knowledge of Lotus Notes. And he seemed to enjoy having me there to witness his cleverness. He was happy to describe what he was doing, and he never minded explaining something twice.

My hard work began when I turned my scribbled notes into a formal Detailed Design document. The PM gave me a template and some example documents, and the level of detail in them was frankly intimidating. AusCM had stringent processes around documentation, with three separate reviews before they could be shown to the customer. In this business, where a lot of the work happened in the depths of computer systems, documents were the key artefacts that proved we were making progress.

It took me several attempts to get the Detailed Design past our reviewers, but their comments were helpful and I learnt a lot about tech writing in the process. I also learnt a lot more about our Notes solution. You need a clear understanding of tech to write accurately about it, and I rang Mr

Notes many times for extra details. By the time my document was done, I understood that solution as well as he did.

It was a good thing I did. A couple of weeks later, just as the PM and I were preparing to leave for a project meeting, he got a call. Mr Notes had been in a car crash and was in hospital. He was expected to make a full recovery, but he would be off work for a while.

I never saw Mr Notes again and I never heard the full story, but I could well imagine what had happened. I'm afraid my immediate concern was not for Mr Notes, but for whether anyone else had been involved.

After relaying this news the PM said, 'Congratulations, you're the tech lead now. You can handle this meeting, right?'

It just so happened that the upcoming meeting was the very one he had warned me about right back at the start. Today, we were seeking formal acceptance of the solution, and the green light to proceed to Implementation. If there were any technical questions, the ones that Mr Notes would usually have answered, they were now on me.

'I can do it,' I said.

I'd seen a few curveballs from this customer, so I was on high alert—but in the end, he only asked for clarification on a couple of points, and I had no trouble answering. At the end of the meeting, the PM and I walked out smiling, acceptance obtained and a significant milestone completed.

*

After all the delays, we were a couple of months behind the original schedule. In our favour now was a promise made by the clone salesman when he'd been at his most desperate.

The servers were to be shipped from the supplier fully built, complete with the Solaris operating system and the Notes server application, installed and configured to the specifications I had sent them. With twenty-odd servers going out to sites around the country, this took a big task off our to-do list—but even with that bonus, the project's profit margin had shrunk to razor thin.

During implementation, I witnessed the sort of logistical operation that AusCM excelled at. A dedicated Implementation Manager took over, planning the hardware deliveries and rollout at each site, including teams of local contractors to run around installing the Notes application on every PC.

I was to be one of three AusCM consultants heading out to the sites. The other two, both men a few years older than me, were married with young kids and didn't want to stray far from home. They split the greater number of sites in the populous south-east of Australia between them, while I took the sites on the far side of the continent.

Compared to these colleagues, I was commitment free and keen to travel. I had recently, and very amicably, parted ways with Lobster—no big drama, just a realisation that we wanted very different things in life, and each cared enough about the other to let them go and find it. There wasn't much happening with the band either. We were all getting older and our priorities were changing.

I would be on the road for five weeks, travelling up the coast of Western Australia, across the top end of the Northern Territory, and down into Queensland. I was looking forward to the trip, but I knew I was in for a tough first week. As well as the rollout of a couple of sites around Perth, I also had another big document to write.

This one was the Operations Guide, a detailed manual explaining how to administer and troubleshoot the specific Notes solution we were installing. Originally, it wasn't due until after implementation, so I'd planned to write it when I returned to Sydney. Now the customer was insisting, entirely reasonably, that because his people had to start administering Notes during the rollout, they needed their documentation *now*.

'It's OK,' I said to the PM, just before I left for Perth, 'it's not like I'll have anything else to do in the evenings. I'll get a full draft to you by the end of the week.'

I made that promise before I saw how he'd skimped on my accommodation. I had been booked into a shabby motel on the outskirts of town—not quite the polished corporate hotel I'd been expecting for my first business trip. But I was tired when I arrived after a five-hour cross-continent flight, and it was enough to have a bed to crawl into.

The following morning, I drove my hire car an hour south to the first site on my schedule. I felt nervous showing up to a place I'd never been. Were they expecting me? Would the server be there? Would anyone be mean like some of the types I'd had run-ins with at Money Bank?

Happily, I had nothing to worry about. A friendly IT guy was waiting for me, and he showed me straight to the server room where the server had already been unpacked and mounted in a rack. When I checked it over, I was pleased to see that the clone salesman, however annoying he was, had been true to his word. Everything I expected was in place.

I had a couple of minor issues with my setup steps, but before long the server was talking to the Notes hub in Sydney, and all my functionality tests had passed. The IT guy took me to the staff cafeteria for lunch, and afterwards I walked him through the solution, showed him how to set up the Notes client on his PC, and discussed plans for the full client rollout the next day. I was enjoying myself: it felt good to be the one in the driver's seat, instead of the spectator I had been earlier in the project.

The day had gone well, even better than I'd hoped, but I felt very drained as I drove back to Perth. I'd been performing the role of Visiting Consultant and keeping up this unfamiliar act had worn me out. All I wanted was to pick up a take-away, go back to my motel, and watch trashy TV until bedtime... but I'd promised the PM I'd get the Operations Guide to him by the end of the week.

Back in my motel room, I forced myself to work. It was hard to get started but I knew that if I kept at it, I'd eventually find my flow and my tiredness would retreat. I'd only just hit the zone when my concentration was shattered by loud rock music and the distinct smell of pot smoke wafting through my open window. I got up and closed it, but that barely muffled the strains of AC/DC taking their long way to the top.

I struggled on for a while, but the focus I'd so briefly found was gone. Annoyed, I went down to the front desk to complain, but the guy there just shrugged and said, 'That's a long-term resident. He'll turn it down eventually.' It went on for hours, long after I'd given up and gone to bed with my head under the pillow.

The next morning, grumpy as hell after too little sleep, I rang the PM and told him to move me to a quieter hotel or I was getting on the next plane home. I understood the financial straits the project was in, but it wasn't on me to pay for it. He was a decent fellow, and I think he felt embarrassed about subjecting me to a place like that. He arranged the new booking, and there were no more bargain basement dives for the rest of the trip.

I'm proud of my twenty-seven-year-old self for taking that stand, but at the same time, I know that if I hadn't had the 'excuse' of a whole lot of extra work I would have quietly put up with the dodgy accommodation. As a young woman travelling alone, I was vulnerable to worse things than a noisy neighbour, but I had a relentless need to prove to the world that I was as tough and capable as any of the guys. Learning how to speak up for yourself when you didn't have a voice growing up is a process, and I had barely started working my way through it then.

At the end of the first week, I had completed two sites and the Operations Guide, sending the document to the PM on Sunday evening. It had been a crazy time. I had effectively worked two full-time jobs, played a new role, met a lot of people, and navigated through unfamiliar places. I was tired, but the over-achievement drive was strong in me then and I was proud of myself for doing everything I'd set out to do.

After that, the trip settled into a comfortable routine. I travelled to a new site, set up the server, briefed the roll-out contractors, and got it all wrapped up well within the allotted time.

My evenings and weekends were now freed up for some fun. I went out on the town with a friend who lived in Perth, and another one whom I ran into by chance in Darwin. I watched the seaward sunsets on the west coast, hiked in the Litchfield National Park in the Northern Territory, and went diving on the Great Barrier Reef off the coast of Queensland.

I even fit in a little something for my indie rock side when I caught US punk legends Fugazi on my last night in Darwin. The next day I spotted them in the departure lounge at the airport and went over to thank them for the show. Lovely chaps that they were, they insisted I join them. I was a little flummoxed, and I can't remember a single thing we spoke about, but it was a uniquely enjoyable way to kill an hour before a flight.

*

When I returned to Sydney, the project was pretty much over, with just a few bits and pieces to complete in the Handover phase. Considering the delays, we had done very well to finish the job only a couple of months past the originally planned year.

'We didn't lose money,' said the PM when I asked him, 'but we didn't make any either.'

Financially, the project had not been a great success for AusCM, but it had definitely been one for me. I'd learnt enormous amounts, I'd seen and done countless new things, and I'd helped build a nation-wide, business-critical system that was now in full-scale use.

My biggest takeaways from the year concerned the business of commercial IT. As well as the Notes project, I had contributed to a few bid responses, so I now had first-hand experience of every stage of a commercial project, from proposal through to completion.

It's easy to overlay the opinions I hold today onto what I thought back then. I could say, for instance, that I left the

Notes project with the conviction that Waterfall, fixed price, and IT are fundamentally incompatible—but it had only been one project, and I hadn't yet reached that conclusion.

I did have some concerns though. I'd seen how much 'finger in the air' estimation went into costing a project. I'd seen a twenty percent markup added for 'risk' and then removed by salespeople incentivised on 'winning the business' but not on final profitability. And I'd been the over-stretched project resource, making up for the shortfall between guesstimation and reality with my unpaid overtime.

Regardless, I ended the project raring to go for another. I had loved being part of a team working towards a shared goal—it was much like being in a band, with everyone having their part to play, and together creating something that didn't exist before. I enjoyed the in-the-trenches humour and the closeness with colleagues that came from fighting through it together. And I really loved the feeling of satisfaction I got from submitting a completed document, from ticking off a milestone, and from watching the solution running perfectly when it finally went live. Those were my payoffs and they made the frustrations, worries, and periods of overwork feel worthwhile.

I believed senior colleagues who said that the problems on the Notes project could have been avoided if we'd simply itemised all the micro steps, estimated the effort correctly, thought through all the risks, and fully documented every possible assumption and exclusion. Next time, if we just did that, profit was inevitable.

These days, I have many, many IT projects under my belt and I have seen far too much to believe in that business model. Apart from the smallest and most straightforward of projects, you just can't know everything in advance when it comes to IT. There is always a new product or version, an unusual technology combination, a unique business process, or something strange in the environment that even the in-house people were unaware of until the project uncovered it. None of that reflects a failure to plan, it is just the reality of IT. And it's a reality that so far, neither the cloud nor AI nor any other over-hyped technological innovation has changed.

There is, however, one thing I still believe from back then, and it's something I learnt on my first day at AusCM. For me, it is much better to be a revenue generator than a cost centre. I need to feel I'm earning my keep, and I need to feel I'm valued. When it comes to working in IT, I still prefer to be the business.

13. How to pivot

I'M NOT SURE WHEN THE WORD 'PIVOT' CAME TO mean a change of direction in a business sense, but whenever it was, it quickly became the single most face-saving word we have in IT. Project hit a roadblock? Reality failed to live up to your expectations? Then just *pivot*! Sounds kind of fun, right?

Had I known this meaning of 'pivot' in the late nineties, I may have worn it out. I executed such major pivots in 1998 and 1999 that by the end of it, my world was literally turned upside-down.

It was all Lew's fault really. In early 1998, when I'd been at AusCM for a very busy fifteen months, he called me into his office. He thanked me for the good work and the overtime I'd done on my projects, and the business I'd help win as part of bid teams. Other than my inability to sugar-coat technical realities when talking to customers, Lew said I was doing a great job.

'I'd like to give you something,' he said, 'but that bean counter of a CFO doesn't believe in bonuses. So instead, I'm going to send you to a conference. Pick one anywhere you like, and we'll pay fees, travel, and accommodation. Just make sure it has something to do with IT.'

Wow! I had never got any kind of bonus before and this one was, as far as I was concerned, way better than money—it was travel! I spent the next few weeks combing lists of tech conferences in far-flung locations, before choosing the Internet Summit in Geneva, Switzerland. I hadn't taken any leave in my first year at AusCM and now I planned to save it for the trip and make it a proper holiday.

I was in the middle of planning my post-conference travels when an email arrived from my old uni friend, and *Fred The Bison* co-fiddler, Diana.

Diana had been living in the States, and I hadn't seen her in years. Her email said that if I could get myself to London in early August, I was invited to a 30th birthday celebration that she was planning with her old uni flatmate, and sharer of her date of birth, Nathaniel. By complete chance, the party was in my final week in Europe.

'I'll be there!' I emailed back.

I was flat out over the next few months, juggling two projects and the occasional bid, but eventually the calendar ticked over to July 1998, and it was almost time to go.

On my last day of work, just as I was about to leave the office, a colleague joked, 'Don't do that thing, you know, where people go all the way to the other side of the planet and fall in love with someone they grew up two streets away from.'

Ha ha, yeah, like that would happen.

*

The conference was OK, some interesting talks, but internet governance wasn't really my area, and I'd only picked this one because of its timing in the European summer. So ignorant was I about the subject matter that when someone mentioned that Tim Berners-Lee was just up the road at CERN but hadn't graced us with his presence, I didn't even know they were talking about the man who had invented the web.

I had arranged to meet up with an old friend after the conference for a trip around France. We hired a car and headed south, spending the next two weeks driving from town to town, soaking up the sun, the sights, and the scents of Provence in the summer. In the cool of the mornings we walked around grand houses and Roman ruins. We searched out rustic town squares where we sat in the dappled shade and lunched on goat cheese salad, washed down with carafes of chilled rosé. The afternoons were hot and sleepy, and we relaxed until the day cooled into approaching dusk. Then we went out, looking for the most interesting restaurant or bar in which to while away the evening. It was a fabulous time, and I couldn't remember when I'd last felt so relaxed.

We passed through Toulouse where I met up with a university friend who was living there. She mentioned the upcoming party in London, to which she had also been invited but could not attend. 'Did you hear that Nathaniel broke up with...', she said, mentioning a name I didn't know.

I shrugged, no I hadn't. I had barely known Nathaniel at university and had no idea what he'd been doing since then. It was his party too, but I was going because Diana had invited me.

*

I arrived in London on the day of the party, tired from travel and an uncomfortable final night in a cheap hotel in Paris. I was staying with another friend in London, and it was so pleasant sitting in her kitchen, catching up over dinner and a bottle of wine, that I contemplated staying in. But I'd come all this way, so I roused myself and headed out, catching the tube to the river-side suburb of Hammersmith.

Dusk was closing in as I walked along the towpath beside the Thames, looking for the rowing club where the event was taking place. As I approached, I could see a man standing alone on the balcony, the music and laughter of a party in full swing drifting out through the open door behind him. Our eyes met and I recognised Nathaniel. He looked very different from the scrawny youth I remembered from uni. He'd grown up, filled out a bit, cut his hair, and hmmm... he looked pretty good. He also looked pleased to see me. He disappeared from the balcony and, moments later, reappeared at the ground-level door where he welcomed me with a hug.

The evening was great fun, and I was glad I'd got myself there. There were people I hadn't seen in years, some currently living in the UK, some, like me, just passing through. It was wonderful to see Diana who had not changed a bit and was still her loud and lively self. Nathaniel and I had a bit of a dance and there was an undeniable frisson, but he was the birthday boy and in high demand, and apart from that one dance, we had little opportunity to chat.

At the end of the night when I went to say goodbye to Nathaniel, I found that he had something else in mind. 'Come home with me,' he said with disarming and, to be fair, quite inebriated bluntness.

I was tempted. I had been deliberately celibate for a year, determined to not compromise on romance. After breaking up with Lobster, I had promised myself there would be no return to one-night stands or friends with benefits, and that anyone I dated had to be a serious contender for The One.

But it had been a while... Perhaps a complication-free fling on the far side of the planet wasn't breaking my rules too much.

Well, except for one complication. 'Isn't Diana staying with you?' I said. 'And half a dozen other people at this party?' Having a fling was one thing, having all those bystanders was quite another. Instead, I agreed to meet Nathaniel for dinner in a couple of days, after his guests had departed.

When the time came, it felt strange to be going on an actual date. Usually, I met guys at the pub or a party or in the band I played in (never at work—I had strict self-imposed rules about that), and then we sort of drunkenly fell into bed at some point. Now here I was, completely sober, catching the tube to the South Bank for a very grown-up sounding stroll along the river followed by dinner.

At the restaurant, Nathaniel and I talked non-stop. I knew almost nothing about him and was fascinated to learn about his childhood split between England, France, and Australia. After university in Sydney, he'd done a master's in political science in France, and had then moved to London where he worked as a policy analyst at the Australian High Commission. The more interesting things he said, the more attracted to him I felt. Ooh yeah baby, talk to me about the influence of European inter-state migration on the geopolitical climate! So hot!

After dinner we walked onto a bridge to admire the river. He kissed me, and any lingering resistance melted away. This time, when he asked me to come home with him, I said yes.

I had only three more days before my flight to Sydney, and Nathaniel had to work, but around that, we saw as much of each other as possible. I was falling for him, hard, but I still didn't think a relationship was on the cards. His future was in London. He was going to start a huge new job the following year as one of the British Foreign Office's 1999 intake of new diplomats. I couldn't ask him to give up that for me—but it felt insane to even consider leaving my whole life in Sydney for him when we'd only been together a couple of days.

On the last night we walked to the top of Primrose Hill to watch the sunset. He held me tightly and I could feel his heart beating. His face, already so dear to me, filled with emotion.

'You know,' he said, 'my contract is about to end, and I was thinking about traveling for a few months before I start at the Foreign Office. What if I just came to Sydney?'

I looked at him, astonished. Did this holiday fling have legs?

Little did I know, Nathaniel's quick change of plans was the first of many pivots to come.

*

Not even two months later, I woke up feeling sick. Nathaniel had arrived in Sydney a few weeks earlier, and we'd picked up our whirlwind romance right where we'd left off in London. What with all the romantic dinners and catch-ups with friends and me showing him around my band pub haunts, there had been a lot of late nights followed by full working days. Perhaps I'd just been over-doing it... but something felt different.

'You poor thing,' said Nathaniel, when I told him I didn't feel well enough to go to work, 'can I get you something?'

'Yeah,' I replied, 'a pregnancy test.'

For years, I had insisted that I didn't want children, too terrified of repeating the mistakes of my parents. Lobster and I had always been on the same page on this point, but as I healed, and learnt to trust my own instincts, my views changed. 'Perhaps,' I thought one day, 'having a child wouldn't be the worst thing that could happen.'

I realised then that I wanted the door into parenthood to be open, even if I never walked through it. My change of heart was a big reason why Lobster and I had separated.

As I lay in bed, waiting for Nathaniel to come back from the pharmacy, the worry made me even more queasy. I had come around to the idea of parenthood, but I didn't want it right now! Nathaniel and I had been together for, what, *five minutes*?? It was way too soon!

Nathaniel returned with the testing kit, and I shut myself in the bathroom with it. When the blue line appeared, confirming what my body already knew, I was dismayed. My life had been going so well and now this was going to ruin everything, I was sure of it.

I wordlessly showed the result to Nathaniel. 'Let's not jump to any conclusions,' he said. 'I think we should go to a doctor and get it confirmed.'

Later that morning, a doctor confirmed it. I was definitely pregnant.

Afterwards, Nathaniel and I walked back through the park. We had barely spoken since the blue line appeared, each lost in our own thoughts, but now we sat on the grass under a shady tree to talk it over.

I felt despondent. I assumed Nathaniel, like all the other young men I knew, wouldn't want the baby. I assumed he would go back to London and then what would I do? I wanted children when the time was right, but this wasn't it. I thought there was only one choice available to me and, while it's not a choice I would criticise anyone for making, right then the thought of an abortion crushed my heart.

I opened my mouth to speak, but before the words came out Nathaniel said, 'It's your decision, but I think we should have the baby.'

And just like that, a world of possibility opened up. Instead of one lonely path, there were suddenly many I could pivot towards.

'I want the baby too,' I said, emotion welling in my throat—but only for a moment before my practical mind took over.

'How would this work? You're only here for a few more months,' I said.

'Come back to London with me,' suggested Nathaniel. 'You can stay home with the baby for as long as you want, we'll manage.'

In those days, I still often doubted my ability to know what I needed, but on this occasion, I was absolutely sure.

'The thing is,' I said, 'I can't give birth in London. It will have to be here.'

A little voice inside was warning, 'If you have this baby in London, you'll be inviting post-natal depression.' I had built myself a support network of girlfriends in Sydney and I instinctively knew that if I was going to pass through this big scary door into parenthood, I would need them.

Again, Nathaniel showed how readily he could pivot. 'I'll write to the Foreign Office,' he said, 'and ask if I can start in the 2000 intake instead.'

I felt dizzy, my life completely turned around. It was October 1998, the baby was due in June 1999, and I now agreed to move to London with Nathaniel and our child in early 2000.

'So, are we going to get married then?' I wondered.

'Yes! I accept!' said Nathaniel (who still maintains to this day that I proposed).

Nathaniel and I were married in December 1998, only four and a half months after the party in London. To invoke another popular IT term, those months had been a sprint.

*

I didn't tell anyone at AusCM that I was pregnant until after our honeymoon, which happened to coincide with the Christmas-New Year break. Back in the office at the start of January, I went first to Lew.

I was worried that he'd be annoyed about a baby disrupting my availability for work, that he'd regret offering me the conference trip that had started all this, and that he'd think less of me for being so careless, but Lew surprised me. He stood up, came around his desk to where I was shuffling nervously from foot to foot, and he hugged me. There may have even been a little tear in his eye.

I didn't immediately tell Lew about my plan to move to London in a year's time, though I don't think that would have changed his response. He could be tough, expecting his people to work as hard as he did, but he really did want the best for me.

I can't say the same about the next colleague I told.

Nathaniel and I had agreed we should keep our options open, just in case of any more surprises. The Foreign Office had delayed his start date, but we were keeping the possibility of staying in Sydney on the table for now. With that in mind, I went to see the Chief Financial Officer or, as Lew always referred to him, 'that bean counter of a CFO'.

'So, big news, I'm pregnant,' I told the bean counter. In contrast to Lew's heartfelt response, he just stared wordlessly at me, waiting for me to get to the point.

'Right, um, so what I want to know,' I went on, 'is what sort of provision do we have for maternity leave?'

'Three months,' he said.

'Is that on full pay?' I asked.

He frowned at me like I was a complete idiot. 'No,' he said, 'of course not, it's unpaid. The only thing we're required to give you is an equivalent job if you come back *full time* within *three months*. And if you don't, we're not obliged to do anything.'

Well, that was hardly worth changing plans for.

At that time in Australia, parental leave was left to the generosity of individual employers, which meant it only existed in industries with a high percentage of female staff and strong unions. My old school friend Emma had recently returned to her hospital job after six months' paid maternity leave, but I was in a medium-sized private company in a male-dominated and entirely non-unionised sector, and there would be none for me.

I felt a little offended that the CFO didn't think I was worth some kind of offer, but he had just made my next pivot very simple. I would work until the baby came, and then I would resign ahead of our move to London six months later.

That year was a strange one in IT. It was 1999, and almost everyone who knew the slightest thing about computers was roped into mitigation projects for Y2K, aka 'The Millennium Bug'. Except me. I had three projects on the go and was flat-out busy, with regular interstate travel. Growing a human in your own body is tiring enough without a demanding job as well, but I soldiered on until late in the pregnancy when I was so sick on a turbulent flight that I told Lew, 'No more travelling.'

In the final month of my pregnancy, I worked in the office, finishing up documentation. When a customer rang about a failed power supply in a server, and with all the other techies out of the office on Y2K patching, Lew asked if I could go and see to it. Welcoming this break from staring at my screen, I grabbed my toolkit and a replacement part and caught a taxi to the customer's office.

I was met outside by a young man. He was expecting a female techie, but not a heavily pregnant one. As I levered myself out of the taxi, I saw his eyebrows shoot up. As he led me to the server room he said, in a casual sort of way, 'So, ahh, tell me, have you done much of this sort of thing before?'

'Yeah,' I replied, 'eight years of it.'

In the server room, I had to ask him to wheel aside the rack so I could fit myself behind it and get at the back of the server. I must have looked quite comical, my big pregnant body wedged in there, but neither of us were laughing. He hovered around, keeping a close eye on me, while I unscrewed the power supply unit, slid it out, and carefully unplugged its internal connectors. The whole time, I was aware of his presence. I'd had customers looking over my shoulder in an interested way before, but this felt different. There's every chance he was worried I might get stuck back there, or suddenly go into labour, but it felt like mistrust.

Later, back at AusCM, I told a couple of colleagues about him and his casual-not-so-casual question. In those days I staunchly refused to admit that my gender might in any way matter in my working life, but something about that experience had rattled me. I wanted my colleagues to agree that the customer's question had been weird—but instead they found it hilarious.

'You totally missed a trick there,' one of them laughed. 'You should have said it was your first day!'

And they both fell about even more while I stood there completely baffled, thinking, 'Why on earth would I *lie?*'

That day turned out to be my last customer-facing one as an AusCM consultant. A few weeks later our son Oscar was born, and my time with AusCM was over.

*

We welcomed the new millennium from a park with a view of Sydney Harbour Bridge, walking distance from the little house that Nathaniel and I had rented for the year. It was an unseasonably cold night, and we had dressed six-month-old Oscar in a furry onesie that someone had given us in preparation for wintery London, not expecting us to need it in summertime Sydney.

We were joined by my father, his wife, and my sister and brothers, now aged between ten and fifteen. We took a picnic, and some bottles of fizzy stuff, and somehow Oscar was wide awake and cheerful at midnight when the Bridge exploded with giant sprays of colourful fire.

After the fireworks had subsided into wisps of drifting smoke, we walked slowly back to our place. My teenaged sister Claire hung close to me the whole way, asking what I thought London would be like, and whether we'd ever come back. I'd made an effort to be part of my siblings' lives ever since I'd been allowed back into the family, and abandoning Claire at an age when she might really need a big sister was particularly hard.

We still had six weeks until our arrival in London, but we would be moving out of our rented house in a few days. Before we left the southern hemisphere, there were a lot of people I wanted to say goodbye to. We planned to spend some time in Canberra with Nathaniel's parents, and a couple of weeks in New Zealand visiting my relatives. I particularly wanted to see my father's parents, who had always been lovely to me. I suspected, as would turn out to be the case, that this trip would be the last time I saw my grandparents.

We were in the middle of packing up the house and preparing for our travels when Nathaniel got an email from the Foreign Office's HR department. There had been a mix up with his security clearance, and they were going to have to start over when he returned to the UK in February. He should expect it to take around six months, during which time he would neither be employed nor paid.

Until that email arrived, we had expected Nathaniel to start work soon after we arrived in London. I had planned to stay home with Oscar for a while, getting used to London life, and giving myself plenty of time to consider work and childcare options. Now all those plans were thrown in the air.

In Sydney, I had already been home with the baby for six months while Nathaniel kept us afloat with the short-term job he'd found. It had not been an easy time for me, and I felt very ready for my next pivot.

'We'll switch,' I said to Nathaniel. 'When we get to London, you stay home with Oscar, and I'll find a job.'

14. How to be on trend

I HAVE NEVER BEEN A TRENDY PERSON. I NEVER wore the latest fashion or the cool sunglasses or got invited to the right parties, and I didn't want to either. To me, striving to be 'on trend' looked shallow and conceited, as well as a waste of time, energy, and money. But I guess I would say that, because I've never been trendy.

Except, that is, for one time in my life. In the year 2000, my first year in London, I got a job in the trendiest sector going, with people for whom trends were everything.

We arrived in London on February 14, a cold, grey and, what with the jetlag and the over-tired baby, thoroughly unromantic Valentine's Day. I gave myself a week to settle in and then, with no computer in our flat, I began my daily pilgrimage to an internet café to scan job listings and send off CVs.

I assumed I'd find something in no time. Early 2000 was the peak of the dot-com boom, and tech skills were more in demand than ever. Like the legendary Dick Whittington who came to London expecting to find the streets paved with gold, I came expecting to find tech jobs lying all over the place, just waiting to be picked up.

At first, this assumption led me to be selective. In my mind, I had 'moved on' from the support and system administration I'd done in the School of Electrical Engineering. I wanted to continue working in commercial IT, *being the business*. At AusCM, we had often hired contractors to help with project delivery, and I figured, at the very least, I could pick up something like that.

Meanwhile, we were living off savings and that was stirring up my anxieties about money. It didn't help that London was *so expensive*, with staples costing almost double what they did in Australia. Until there was money coming in, I couldn't even start to settle into my new life.

After three weeks of fruitless searching, I had to admit that the project contractor idea was going nowhere. I was reluctant to return to the front line of IT, to go back to dealing with users' frustrations and helplessness, but already any job was looking better than none. 'There's nothing wrong,' I reassured myself, 'with going back to your trade.'

The next day, I submitted a few applications for sys admin roles, and my luck changed immediately. Within an hour of returning home, the phone rang. It was a recruitment agent, and he sounded almost breathless with excitement.

'I can't believe this,' he said, 'you're perfect!'

Well now, wasn't that nice. I assumed he must have been impressed with my extensive list of work experiences and certifications and wondered which of them had made me stand out.

He quickly outlined the job, which was nothing more than a general IT admin and support role for a new company. It didn't sound special, but he seemed to think so.

'They're going to love you!' he gushed. 'I've been trying to find someone for this client for a while, but it's been difficult as they're *very particular.*'

It was all a little odd, but at least I had a solid lead now. I figured I'd find out sooner or later what was so *particular* about this employer, and what particularly made me their perfect match.

The agent arranged an interview, and only a day or two later I caught the tube to Notting Hill. Following his directions, I was surprised to find myself in a residential street. I double checked, and yes, it was definitely one of those grand houses, all stately and white, that lined the road.

The next surprise was the cop standing at the front gate. Worried that there was some kind of incident, I said, 'I'm looking for an office, Ready2. Have I got the right place? Is it OK for me to go in?' He nodded, stepped out of my way, and pointed to the door of the basement flat.

A young woman opened the door. As she welcomed me in, I asked, 'What's with the policeman at the gate?'

'Peter Mandelson lives upstairs,' she told me.

The name rang a bell as someone in the Blair government, but Nathaniel had to fill in the rest for me after I got home. Peter Mandelson was then the Secretary of State for Northern Ireland, a potentially dangerous position considering the history of IRA bombings. Later, when I worked there, I never saw Mandelson, but I quickly became accustomed to the constant presence of a bobby at the gate. When it rained, they'd shelter at our doorway and we'd offer them cups of tea.

I was ushered into a room which had a desk at one end but otherwise looked nothing like an office. Its main feature was a clothes rack, weighed down with an assortment of brightly coloured garments. There were shelves crammed with shoes, piles of accessories, and a full-length mirror. Apparently, I was going to be interviewed in a dressing room.

Two women were sitting on opposite sides of the desk, and they stood up as I entered and introduced themselves as Carla and Irini. Then they both squeezed in next to each other and waved me towards the chair opposite.

Carla started with some background. Ready2 had been founded by two women named Trinny Woodall and Susannah Constantine... Here Carla paused and looked at me expectantly, but the names meant nothing to me, so after a beat, she went on. Trinny and Susannah wrote a popular fashion column in the Daily Telegraph, but their ambitions went far beyond that. They had obtained over £5 million in venture capital to expand their vision onto the World Wide Web.

'We're starting with fashion,' said Carla, 'but there are big plans to extend into all areas of life that are important to women. It's going to be huge.'

Carla asked me a few simple questions, essentially confirming what was in my CV. At the end of the interview, she and Irini nodded at each other and Carla said, 'We think you're perfect. We'll talk to the Managing Director, and you can expect a call.'

I left feeling very pleased. The location of the office was unconventional, but the dot-com ethos was all about unconventional, and who was I to argue with that? I was still none the wiser about why the agent had such difficulties recruiting for a standard IT role, but I didn't dwell on it. The important thing was that someone wanted me.

After I worked at Ready2 for a while, I realised what had made me so *particular*. You're probably way ahead of me, but I can be chronically unobservant at times.

Trinny and Susanah wanted their website for women to be entirely staffed by women. 'All our employees are female,' they'd say, '*even our IT person*.' I wasn't hired for my CV; I was hired for my gender.

Once I figured that out, I didn't like it one bit. I felt like a rare accessory, valued only for my exclusivity and not my real worth. I was something to show off, I was a fancy fricking handbag.

But that understanding came later. On my first day on the job, all I was thinking about was their IT.

MONDAY 13TH MARCH 2000, LONDON

I wouldn't call this a desk–more a shelf. There is barely enough room for my monitor and keyboard, and with the PC chassis on the floor below, there's nowhere for my feet.

It's my first day and I've been put in the dressing room with Carla and Irini. They sit behind me on opposite sides of the desk, their back-to-back monitors crowding the workspace. It's going to be a nuisance whenever Carla wants to get in or out–I'll have to stand and move my chair out of her way.

This was the only spot they could find for me as the other rooms in this basement flat are as full of desks and chairs as they're going to get. Network and power cables are haphazardly looped over doors and furniture, and run treacherously along the floors. Almost thirty women are crowded in, all sitting at some sort of desk and tapping away at PCs, doing who knows what. Some have job titles I recognise: Accountant, Human Resources, Managing Director. Others were introduced as 'Fashion' or 'Accessories' or 'Cosmetics', so I guess they're producing content for the website.

The only person with any kind of computer background is Irini, a programmer.

'Are you developing the website?' I ask her.

'No, there's another company doing that—I was hired recently to make sure they're doing it properly.'

This seems strange. When your entire product is a website, shouldn't you be developing it yourself and not trusting it to another company? But this is only my first day in dot-com land, what do I know?

I haven't yet met Trinny and Susannah, though I saw plenty of them on my tube ride this morning. Hot pink posters are plastered along the station walls picturing my new bosses, naked from the waist up, Susannah holding melons in front of her breasts, and Trinny delicately positioning a pair of fried eggs in front of hers. They've clearly got a talent for marketing—you can't miss them.

I'm only a couple of hours into the job and I'm worried already. I'm supposed to be planning the computer and phone fit-out of the new office ahead of the company relocating in a few weeks, but there are so many problems with their current IT that need my attention.

Up until now, IT support has been done by a couple of freelancers who turn up on motorbikes when summoned. I spoke to one of them on the phone this morning to get the background.

'Look, I know it's all a bit of a mess there,' he said, 'but we were originally asked to set up an office for twelve people—we didn't plan for so many!'

He gave me the admin password for the office server, along with a warning. 'It's very overloaded and we haven't really had the time for proper maintenance. I suggest you get onto that first!'

He was right, the server's in a poor state. The disk is full, the logs are rife with critical errors, and the virus scanner is spitting out an alarming list of infections.

I'm fully occupied tut-tutting over the server, when the door flies open and in rushes Trinny Woodall. Without glancing towards us, she flings off her clothes, stripping down to her knickers, giving me even more of an eyeful than the tube commuters are getting. I look at Carla and Irini, but they appear completely unperturbed, working away as before. Meanwhile, Trinny has grabbed clothes off the rack and is pulling them on. She burrows through the pile of shoes, selects a pair and runs out the door, the shoes dangling from her hand.

'What on earth was that?' I ask my office mates. Carla shrugs, 'She's probably running late for a photo shoot'.

Clearly, this is not going to be like anywhere I've worked before.

Just before lunchtime, the Managing Director comes around to say hello. She tells me there's an all-company lunch today, 'But I'm sorry, as you're the newest person, you have to stay and answer the phone'. I'm not bothered–I have a lot of poking around still to do.

After everyone leaves, I take my first look at the website, Ready2Shop.com. It's bright pink, just like the posters in the tube. The site is supposed to give personalised tips on clothes, currently available in chain fashion stores, that will suit your shape and preferences. I like the sound of this. I'm hopeless at buying clothes and I consider change-room mirrors to be a form of torture. Anything that lessens that ordeal is welcome.

I soon see that Ready2Shop.com won't give me advice until I register. I start the process and, for the third time today, am confronted with an almost-naked

Trinny Woodall. An image of her body fills the screen with clickable markers pointing to her arms, breasts, thighs, hips, and ankles. I am directed to click through these markers, indicating whether I love or hate that part of my own body. So, just to be clear, I am supposed to compare my eight months postpartum body with the statuesque and model-thin Trinny, and find a single thing I don't immediately loathe about myself?

I am sitting at my shelf, dejected about my hateful body, when all the phones in the basement flat ring together. I get up and answer the nearest one. 'Hi, um, hello, Ready2?'

'I don't understand how to buy anything on this site,' says an irritated woman. 'I've gone through and picked the things I want, and they're in my list, but there's no way to buy them!' Well, I think to myself, at least she got through the registration.

'I'm so sorry,' I say. 'This is my first day here and everyone is out. Can I get someone to ring you back?'

Later, when they all return, I hand a note with the woman's number to one of the staff and tell her about the call.

'Humpf,' she humpfs, 'we get that all the time. They don't understand the *concept*.' As she heads back to her desk, she screws up the note and tosses it into a bin. I am perplexed. Whatever this *concept* is, I don't understand it either.

*

Two weeks into the job, family life had settled into a new routine. Always the early riser, I got up with Oscar when he woke at the crack of dawn, letting Nathaniel sleep a bit longer.

Oscar, at eight months old, had learnt how to drag himself commando-style across the floor and how to get into a sitting position, though he still often toppled over. He was so cute that it wrenched my heart, a twisting feeling that tightened my throat and pricked at the corners of my eyes. He was simply extraordinary, this little creature that I had grown in my own body. I still sometimes felt a sideways jerk of unreality at the impossible idea that I was a parent, despite the proof in the red lace of stretch marks across my belly and the very real presence of this little person, now looking thoughtful as he pursed his lips and attempted to blow a raspberry.

After a little while, Nathaniel emerged from our bedroom, made himself a cup of tea, and took over baby supervision duties. I showered and dressed for work, and then I kissed them both goodbye and left. I loved them with a whole-hearted intensity, but I couldn't deny how giddyingly free it felt to leave the flat alone, unencumbered by bags and a pushchair and the exhausting vigilance of caring for a baby.

As I caught the tube to Notting Hill, I reflected that my first two weeks at Ready2 had passed in a flash. There had been no gentle settling in period, and I had been flat out from day one dealing with an endless stream of tech demands and issues in the basement flat. I'd barely had any time to plan for the office move.

I saw the new office for the first time later that morning. Trinny, Susannah, and the Managing Director were going there to catch up with the decorator, and I tagged along to take a look.

It was just down the road in Westbourne Grove. The office took up the whole level above a supermarket in one big, open space. On walking in, the first thing I noticed was the floor which had been entirely tiled with silver panels.

'Oooh, they look lovely,' Susannah gushed as we all stopped to admire the shiny floor stretching away across the empty expanse.

The decorator preened, 'They were very hard to source but I think the effect is well worth it, don't you?'

I was more interested in what was under the panels. The decorator levered one up to show me the power and data cables laid beneath, with outlet ports neatly cut out at regular intervals. 'Where do the cables terminate?' I asked, and she directed me towards a small room in the corner, little more than a cupboard.

This was to be my server room. It had a rack which housed only a patch panel from which the data cables dropped in zip-tied bunches and disappeared into the floor.

The Managing Director came to the door. 'Does it all look OK?' she asked. 'There's more equipment coming—an office phone system and a new server and some other things our computer supplier said we needed... I'm not exactly sure what. You can check the purchase order.'

'Just one thing,' I said, 'there's no air-conditioning in here. It's a small space and the equipment is going to generate a lot of heat. We have to keep the temperature down to avoid the servers overheating and malfunctioning.'

The MD frowned. 'Well I suppose, if you have to, but it will have to be a cheap, portable one. We're already way over budget on this fit out.'

I wondered then, was the company facing financial difficulties? You wouldn't have thought so from watching Trinny and Susannah as they happily discussed the office layout with the decorator. 'And we'll put Fashion along here, and Graphics up the back there, and our desk right in the middle here. And down in this corner, this is the perfect place for the photo studio, we'll just need some partitions built ...' I only noticed then that the more excited Trinny and Susannah became, the more worried the MD looked.

Back in the basement flat, I asked our accountant for the purchase order and, while I was at it, a list of all IT licenses and equipment we already owned. She apologised, saying that the accounts had been a mess when she started a couple of months earlier, and the best she could do was let me look through the receipts.

I returned to my shelf with several folders and pulled out anything tech related. Among receipts for computers and payments to the motorbike guys, I found that we owned at least fifty Ready2 domain names, purchased in batches over the last year. Carla had mentioned plans that went way beyond clothes shopping, and I could see them spelled out in domains that included Ready2Decorate.com, Ready2Work.com and Ready2Travel.com. In the most recent batch, I kid you not, was Ready2Die.com.

After reviewing this pile of paperwork, I went back to the accountant. 'I've found receipts for four Apple Macintosh G3s but I haven't seen those,' I said.

'Oh yes, that will be the Graphics Girls, they're in the other flat.'

There was another flat?

She gave me the address and, later that afternoon, I popped around to take a look. This flat was much smaller than ours, just a studio, but also less cluttered and chaotic. It housed only four young women, their colourful, large-screen G3s, and a makeshift photographic studio consisting of a couple of lights on stands and a purple backdrop.

One of the women walked me through what they did. The Fashion Girls turned up with clothes borrowed from high street stores, which they modelled for photographing. The Graphics Girls selected, cropped, and colour-corrected the photos, then uploaded them to the site. Lastly, the Fashion Girls tagged each photo with its style, suitable occasions and, the key metric for the Ready2Shop website, the body shapes it would suit.

I should explain at this point that my use of the word 'Girls' to describe fully grown women is true to Ready2 parlance. Trinny and Susanah were routinely referred to as 'The Girls'—something that confused me at first as they were in their late thirties and didn't look like girls to me.

My visit with the Graphics Girls showed me just how labour intensive it was to produce content for the website. And the ridiculous thing about that was that the site didn't even work properly.

I had eventually gritted my teeth and got through the site's registration process—partly in the spirit of 'know your product', but mostly because I wanted recommendations for trousers. The 'you can never be too thin' nineties had not been easy on a pear-shaped woman like me. I hated wearing skirts and dresses which have always made me feel vulnerable, but shopping for trousers was an exercise in misery. They stretched and pouched in all the wrong places, unflatteringly tight across what women's magazines called my 'saddlebag' thighs, while at the same time uncomfortably loose around my waist.

Thus it was with high hopes that I entered my search criteria, only to have them quickly dashed. Every single pair of pants the site offered was explicitly marked as 'No good for the larger behind'. Sigh.

'The site doesn't even work,' I said to Irini.

'Don't worry, I know,' she said. 'When I started here the developers were refusing to do any more work on it because Trinny and Susannah kept changing their minds. I've had to make them promise to stay out of it while I push the developers to fix what we've already got.'

We both shook our heads and sighed about the stupidity of people who thought original code could be produced as easily as snapping your fingers.

*

A month into the job, my list of gripes was growing. People yelled out my name all the time, but it was almost impossible to reach them without tripping over cables or handbags or dogs. That's right, dogs. If you weren't lucky enough to have a female IT techie as a fashion accessory in year 2000 London, the next best thing was a rescue from the Battersea Dogs Home. Trinny and Susannah had adopted one each and regularly brought them into work. They were well behaved enough but they got underfoot, and they contributed to the general funk in the flat. To be fair to the dogs though, the worst culprit when it came to the smell was the Cabbage Soup Diet that half the women were on.

I had liked the idea of working in a company of women, but I had little in common with them. Most of the 'girls' were passionately interested in brands and labels, while I turned up to work in generic clothing and comfortable shoes. I might have known how to fix a jammed printer, but I didn't get their cultural references at all. I caused ripples of incredulity on the day Trinny and Susannah dragged up as 'Posh and Becks' for their weekly Telegraph column, and I had no idea they were impersonating a Spice Girl and her footballer husband. In that all-female company, I felt more like a fish out of water than I ever had in a roomful of computer guys.

I was also unaccustomed to the ways of the British upper class. Both Trinny and Susannah came from families that boasted knighthoods and country estates in their ancestries. They had gone to fancy schools, though I was appalled to learn that Trinny had boarded at hers from the age of six. She's on record saying that gave her a hard edge, and I can attest to that. I came up against it more than once.

Like one day in the basement flat when I was going about my business, and Trinny called out across the room, 'Carol, if my Palm Pilot doesn't work properly in the new office, you're fired!' I stopped and looked at her in astonishment and she immediately doubled down: 'I mean it, I will fire you!'

It felt like a slap in the face. No one had ever threatened me with the sack before, and here she was doing it over a Palm Pilot I hadn't even known about! I went back to my shelf and sat there for a while, staring blankly at my computer screen, too upset to work. As far as I could see I had two choices: pretend nothing had happened or confront her.

And damn it, I was sick and tired of pretending that bullying didn't hurt!

I went back, picked my way around to Trinny's side, and asked if I could speak to her. What I then *meant* to do was calmly tell her that I didn't appreciate being threatened. Instead, what I *actually* did was burst into tears and run to the loo. Stupid emotions!

By the time I got back to my shelf there was an email from Trinny. She apologised, claiming it had been a 'joke'. The email also implied that this sort of thing had happened before with other employees who were equally unused to such witty japes. I thought she was a coward for not coming and speaking to me personally, but an email was better than nothing.

Thank goodness for Irini—she was the one person there who thought like I did. She understood the problems with the website, and she knew there was a lot more to building a good product than waving your hands about and having fabulous ideas. She also seemed to be the only one who wasn't on a diet. Around lunchtime she'd say to me, 'Come on, I'm starving!' and we'd go to the supermarket and buy a roast chicken and bread and salad, and then come back and sit together in the kitchen, sharing our feast while everyone else microwaved their bowls of stinky soup.

*

At the end of each day I caught the Tube home, frustrated by the regular delays as I couldn't wait to see my darlings. Nathaniel was quick to hand the baby to me, especially when Oscar was grisly and whiny. I gave Oscar his bath and got him ready for bed, and then I sat by his crib, reading to him until he settled down to sleep.

Once Oscar was safely tucked away, Nathaniel and I had our time together. We opened a bottle of wine and cooked dinner and talked about our day. Nathaniel told me about a cute new thing Oscar could do, and I told him about the latest bizarre happening at Ready2.

Those were the good evenings, but anyone with a baby knows you can't count on them. Sometimes Oscar wouldn't settle, and we didn't get our much-needed couple time. Sometimes we were both stressed and wanting sympathy at the same time.

We had a nasty fight in my early weeks at Ready2. I came home grumpy, perhaps because Trinny had been smoking in the office again and the smell made me nauseous, or perhaps because the constant demands for tech support made it impossible to plan for the looming office move. I wanted a sympathetic ear for my complaints, but instead I found my usually calm and accommodating Nathaniel stressed and annoyed himself. Oscar had started pulling himself up on the furniture, his little hands reaching for things previously out of reach. Nathaniel had, that day, narrowly averted a potential catastrophe involving a full mug of hot tea.

'Don't complain to me,' he snapped when I started doing exactly that. 'This is what you wanted. You love it.'

How UNFAIR!!

Later, after the shouting was done and I'd calmed down, I realised, with a queasy sense of shame, that I had been the asshole here. It's not just men who fall into the easy trap of thinking that workplace busyness is more important than the daily labour of caring at home. I was disappointed to realise that I had done exactly that myself.

It was the reality check I needed. Since then, I'm proud to say, Nathaniel and I have maintained an equal opportunities partnership in which we value and appreciate each other's contributions. While our kids were growing up, we did all the juggling combinations: one working, both working, one working part time, one or both working from home, hired help, no hired help. We have shared in the raising of our children while also pursuing our careers, and I am proud to say that we never needed to have that particular argument again.

*

I certainly needed Nathaniel's patient forbearance during the weekend of the office move which arrived at the end of April, seven weeks after I started at Ready2. The staff were given Monday off so I'd have three days to get everything set up, but it was going to be a push.

I had already put in long hours over the previous weeks, installing and setting up the phone system, building the new server, and designing a standard build for the PCs. Now I was going to have to transfer all the data from the old server to the new one, rebuild around forty PCs, reconfigure the Macs, and get the printers set up on the new network. It was too much for me alone, so the MD had hired a contractor to help over the weekend. But even with the extra pair of hands, there was so much to do that I only went home to sleep.

Trinny and Susannah wandered into the office on Monday afternoon to check that everything was ready for the start of business the next day. They were disappointed to see that the silver floor had all but vanished under the desks, chairs, shelves, and cupboards that now filled the space.

'Why do the computers have to have so many *cables*,' sighed Susannah. 'They look so *ugly*.'

If they thanked me for my efforts over the weekend, I did not record it in the journal I was keeping at that time.

At least the staff were happy when they arrived on Tuesday morning. Finally, we were all in one place, and it was a lot less cramped than the flats. I hoped that many of the tech troubles would disappear now I had a decent server, proper cabling, and a uniform setup, but I didn't get any less busy. People still shouted out my name all the time, and it was as stressful as my early days in the School of Electrical Engineering. Just as I'd done back then, I wrote a to-do list of improvements that should help, but in Ready2's open plan office, I couldn't get far enough away from my users to implement any of them.

One morning, I came in to find a boxed Apple laptop on my desk with a note on it from Trinny saying it was a present for her father and instructing me to set it up. I picked up the box, marched straight over to the Managing Director, and dropped it on her desk.

'This is a disgrace,' I said. 'I work for this company, not the Woodall family.'

The MD read the note and shook her head. 'Don't worry,' she said, 'I'll deal with it.'

By this time, I knew I would never see eye to eye with Trinny. I felt that she didn't take my opinions seriously, and I believed it was because she didn't like the way I looked—I'd seen the way she curled her lip in distaste whenever I shambled into view. As a fashion accessory, I was a decidedly dowdy one.

Even so, any views she may have had on my appearance didn't stop her boasting about me. In the new office, my desk wasn't far from hers, and I could hear her talking on the phone to the press or potential investors, using her line that 'All our staff are female, *even our IT person*'.

My proximate position left me with no illusions about the company's financial state. While the Fashion Girls still went out to scour the shops, bringing back borrowed items to be photographed and uploaded, the real business had become the hunt for new investors. And it didn't seem to be going well.

In my first week at Ready2, the NASDAQ trading value for tech stocks dropped twenty-five percent. There had been some minor rallies since, but the overall trend was inexorably down. The dot-com bubble was deflating rapidly, and investors were no longer impressed by a website and a sketchy business plan—they wanted to see genuine profitability. Ready2 didn't even have a revenue stream and this so-called business possessed only outgoings.

The lack of money-making potential had bothered me since my first day when I took that call from the woman who couldn't buy anything on the website. Once I got into the site myself, the source of her confusion became clear: it was laid out like a shopping portal, but all you could do was add items to a list. To buy anything, you had to print out the list and go to the shops yourself. And it wasn't like Ready2 got a kickback if you did—they were effectively advertising the products of high street fashion stores for free.

I asked the Managing Director how Ready2 was ever going to make money, and she gave me the business plan to read. It included a lot about how Trinny and Susannah were going to improve the lives of 'time-poor, aspirational women', but nothing about how they proposed to fund it.

'So the idea,' said Trinny when I asked her, 'is that we'll make the site so good that women will give us more and more information about themselves, and then we'll sell it to advertisers.'

Not just 'improving lives' then. While I didn't feel comfortable about selling personal data (and I don't think anyone does now), at least I could see its money-generating potential. But it didn't ring true. If your main goal was for your website to be a honeytrap for personal data, wouldn't you at least make sure the damn thing worked?

I started to suspect that the real goal all along was an overvalued IPO. A year before, they might have succeeded, but midway through 2000, the moment for exorbitant profiteering from dot-com smoke and mirrors had well and truly passed.

*

By mid-June, three months in, I was so sick and tired of never-ending support demands that I considered looking for a new job. But we had just found Oscar a childminder, and Nathaniel would finally be starting with the Foreign Office in a couple of weeks, so it didn't feel like the best time for me to quit. And then Carla came to talk to me about plans to get into 'mobile' and 'interactive TV'—noting that this was before both smartphones and YouTube and I had no idea what she was talking about, if indeed she did. She said they wanted me to be 'Head of New Technologies', which I took with a large grain of salt, until the HR lady also came by to say they were hiring an assistant for me, 'sooner rather than later', so I could focus on this other stuff. I decided to stay a bit longer and see how it panned out.

Perhaps these trendy-sounding ideas came from the woman who had suddenly appeared in the management team. She was tall and loud, rode a motorbike, and supposedly had a kickass reputation for snagging investors. She was fond of pronouncements like, 'We just have to make them show us their knickers before we show them ours!' Then she'd swagger off, leaving Trinny and Susannah repeating her latest kickass soundbite to each other, like it was some kind of mantra that could save their company.

While their hunt for a new investor continued, the spending didn't slow down. I was given a bunch of CVs for my assistant and interviewed the candidates myself. There were no women, so I selected the next best option: a gay guy who thought Trinny and Susannah were absolutely *fabulous*.

My assistant started in mid-July, and I was still getting him up to speed when Kickass disappeared. The official story was that she had an unrefusable offer in Japan, but she never did hook an investor for Ready2, and I think she was getting out before the stench of failure tarnished her precious reputation.

Soon after, in early August, Irini quit. She had a job offer from IBM and told me she couldn't wait to get out of this circus and go work somewhere with grownups. I was seriously considering following her out the door when the Managing Director took me aside for a private conversation.

'We have to make some cuts,' she said, handing me a list of names. 'We're going to take them all out of the office tomorrow to tell them. While we're out, we want you to disable their accounts.'

Two-thirds of the company were on the list. I saw my new assistant's name, but not mine.

'Let him stay,' I said, 'I want to go too.'

An invitation to an 'all staff' lunch at a nearby pub was sent to the whole company, followed by emails to selected people that told them to quietly ignore the invitation. Even though I was now on the redundancy list, I had agreed, as my final task, to stay behind and disable accounts. The irony did not escape me that my five months with Ready2 were bookended by lunches I wasn't invited to, though I certainly didn't want to be at this one.

I watched as thirty young women left the office, chattering happily and gesturing to the few remaining at their desks to hurry up and join them. 'Uh, yeah, I just have to finish this, I'll catch you up in a minute,' said the remainers, and then stayed where they were, pained expressions on their faces.

After the leavers had left, I went into my little room and stood at the server console, disabling accounts. Once I'd finished, I handed over the job of disabling my account to my former assistant, picked up my stuff, and went home.

*

Ready2 limped on for another six months after I left, but the dot-com trend was well and truly over and they had to fold. Trinny told the press they were 'hibernating' the idea and lamented the short-sightedness of investors who couldn't see a clear value proposition in throwing another twenty million quid after the first five.

If the names Trinny and Susannah are familiar to you, you'll know that they survived the failure of Ready2 just fine. Shortly afterwards, they created the television show *What Not to Wear* which became an international hit. These days, Trinny helms her successful *Trinny London* skincare brand which, according to an interview I read, she built slowly with profitability in mind. Like a proper business.

I'm sure the failure of Ready2 was hard on Trinny and Susannah, but they always looked to me like the sort of privileged, connected people who land on their well-supported feet. Susannah once told me that, despite the journalist career she'd had before Ready2, she had never once done a job interview.

Half her luck, I thought, as I again trawled the job sites, looking for my next gig. Whatever it was, I vowed it would not be one with people who cared more about fashion than functionality—I wasn't going to make that mistake again.

15. How to make the same mistake twice

IT WAS A HOT AND STICKY LATE SUMMER'S DAY, the sort that London really doesn't do well. The tube had been airless and oppressive, and I was sweating in my suit as I arrived at an office building somewhere in the city.

The interview was with Bruce, an Australian in his early forties. His office had no aircon and was as uncomfortable as the tube. When Bruce leaned back in his chair, hands interlocked behind his head and elbows splayed out to either side, the damp armpits of his shirt were on full display. 'We're moving out of this shithole in a couple of days,' he said, 'somewhere much better.'

Bruce explained that his company NewzServe was a startup (*oh dear*) but not one of those stupid flash-in-the-pan websites (*OK...*). Bruce told me that his background was in print media, that he'd been Head of IT for a global newspaper conglomerate. That's where he'd come up with the idea of starting an 'Application Service Provider' targeting the print media industry.

'You see,' he explained, 'typesetting and layout applications are expensive. It's bad enough for a big company, let alone the smaller publications. So the idea is that we'll host the applications and pay the licence fees, and the customers can access them over the internet on a time-share basis. They get to use the application, but they only have to pay when they need it. Brilliant, right?'

It is now, in 2025, so commonplace to subscribe to applications 'in the cloud' that many businesses exclusively use software this way. It's like how we now stream music and movies instead of buying copies to play on specialist hardware. Back then, in the year 2000, I still listened to CDs and I still watched DVDs, and I'd never heard of running software across the internet. Considering how slow most internet connections were, it didn't even seem possible—but Bruce insisted that it was.

'I've seen this great technology that will make it work,' he assured me. 'If you get the job, I'll explain it all to you.'

He asked about my background, and I talked about AusCM—how I'd had to learn fast so I could implement large-scale solutions with enterprise software I'd never worked with before. I also told him a bit about Ready2 and how I'd never really seen eye to eye with Trinny and Susannah.

'This city's full of posh idiots like that,' said Bruce. 'It's a different world over here. But I'd never get something like NewzServe off the ground in Australia, that market's too small, so I have to put up with them.' At the end of the interview, he said the job was mine if I wanted it.

I went home and thought about it. I was worried about joining another startup, but I was encouraged by some key differences to Ready2. Firstly, NewzServe's fee-for-service model was easy to understand. Secondly, while Ready2 had been wholly investor funded, NewzServe was started with a combination of venture capital (or 'vulture capital' as Bruce called it) and the lump sum payout Bruce had received when he'd left the news company. He had skin in the game, and I assumed that would make him careful. I was also intrigued by this idea of an application service—it sounded cutting edge, and that was exciting.

I talked it over with Nathaniel, but he said it was up to me. Now that he was working we needed two incomes to pay for Oscar's childminder, but Ready2 had given me a full month's salary in redundancy, so I had a little wiggle room if I wanted to keep looking.

In the end, my decision was largely an emotional one. Since moving to London, I had struggled with homesickness, and I liked Bruce's unapologetic Australianness, sweaty pits and all. I phoned him the next day and accepted his offer.

During the week that it took to sort out my employment contract, Bruce moved the company to a serviced office, high up in the Lloyd's building. 'It's bloody expensive,' he said, 'but the Poms don't trust you unless they see a fancy address.'

The Lloyd's building is a modernist landmark of the City of London. It is a great metal structure, with the lifts and plumbing on the outside, reminiscent of a rocket ship's launch frame. I enjoyed those lifts—I found it exhilarating to watch the street drop away beneath my feet as I shot up the side of the building in a glass and metal box.

On my first morning in the office, I met the only other members of NewzServe: Bruce's business partner and their shared executive assistant. I was the company's second hire.

Now that papers had been signed, Bruce was ready to share more details. The product that had inspired his idea was made by a company called Citrix. According to Bruce, Citrix provided the key ingredient that would allow applications to be used from a different location, even across a slow network connection.

I had questions but Bruce stopped me. 'You'll know all about it soon—I've got you booked on the training.'

In the meantime, I had a simple IT job setting up a server for office internet, email, file and print. 'We've been using the IT services here and they charge like a wounded bull,' said Bruce. 'It will be good to get off that and use our own server.'

Two days later, I started my Citrix training. The first thing I wanted to understand was how applications could possibly work across an internet connection.

'You see,' explained the trainer, 'we're sending very little data across the network—just mouse clicks, keyboard, and screen updates. The application is fully running on the remote server and none of its data crosses the network.'

'So, it's sort of like you're watching a live broadcast on TV?' I said. 'With just the pictures coming to you?'

'Yes, only better, because you can direct the action with your mouse and keyboard,' he confirmed.

I wasn't entirely persuaded, but then, as an exercise, we played with an application running on a server in New York, and it was perfectly usable. After that, I was sold. I returned to the office convinced that Bruce was really onto something.

*

My next task was to prepare a demo for a print media trade fair. It was happening in Amsterdam in a couple of weeks' time, and I was going to accompany Bruce to demonstrate his concept to potential customers. Not yet in a position to afford his own booth, he had arranged for us to squat in the corner of another exhibitor's display. We would show their application running hundreds of kilometres away on a server in our London office.

'They'll see the application running on the computers there,' enthused Bruce, 'and then right next to it, one running from London! And it will be just as fast!'

I ordered a couple of PCs, one to be the Citrix server that would stay in London, and the other to be the demo machine that we'd take to Amsterdam. I installed the vendor's application on the server—I had no idea what it did, but according to Bruce, everyone used it in the world of print media. It's not like I needed to know anyway, I just had to show it running from the other side of the Channel.

The trade fair was quite an experience. I'd been to a few IT conferences by then, and I'd walked through the exhibition halls, but I'd never seen anything like this. Many vendors were showing machinery. One had erected an entire print assembly machine, and it towered above the hall, gleaming silver, looking like an alien spaceship. Machines whirred and spat out printed pages at speed. The noise was continuous, and the air was filled with the dizzying smells of ink and toner.

I saw many software vendors and was amazed by the sheer number of specialist applications for the print industry, none of which I'd ever heard of. The booth that I'd be inhabiting belonged to one of the larger vendors. It was big enough for a coffee bar, different seating areas, multiple demo stations, and piles of branded freebies.

Looking back on the lavish displays at that trade fair, the print media industry expected no serious threat from the internet. It was October 2000, dot-com stocks were continuing to tumble, and every week another big name went down—the internet would have looked to many like an over-hyped failure, and nothing to worry about. It wasn't until many years later that I appreciated the unintended irony in Bruce's plan: he wanted to sell services to print media over the very thing that was going to ruin them.

We had travelled to the fair together but, after dropping me and my computer at the booth, Bruce disappeared. Someone showed me to a small standing table in a corner of their exhibition space. I hung a NewzServe sign on the table, then set about unpacking, assembling, and testing my computer.

I stood there for two days. The experience was disturbingly reminiscent of the Apple job I'd had half a lifetime before. Just like then I sold nothing, although this time, there was even less interest.

People asked about the publishing application, and I had to say, 'I'm sorry, you'll need to talk to one of the vendor people about their application. I'm showing how it can run just as quickly from a computer in London—it's a great new system for cutting down on licensing and server costs. Do you want to hear more about that?' They'd look confused and back off, glancing around for rescue. When anyone did let me demonstrate the future of remote applications, they clearly saw it as little more than a curiosity.

Other than the couple of times he brought someone over to see my demo, I barely saw Bruce until the fair was over. As we travelled back to London, Bruce boasted about the encounters he'd had around the fair. 'People think NewzServe is a great idea,' he crowed. 'They love it!'

I, on the other hand, was already having doubts. During the last two days, while standing awkwardly by my demo computer, I'd had plenty of time to think. One thought that had occurred was, 'Does our host vendor understand that NewzServe might reduce their licensing revenue?'

A big part of Bruce's pitch was licence sharing. Instead of each customer buying their own, NewzServe would buy the bare minimum to cover concurrent users. The way Bruce talked about it, he'd sign up half a dozen customers but only pay for one or two licences. This was a direct revenue challenge to the application vendors, and yet one of them had agreed to us squatting on their booth?

'No, it's fine,' said Bruce when I put this to him on the train back to the UK. 'This will lead to them selling *more* licences. We'll be signing up customers who would never have bought their product in the first place, because it's too expensive for them.'

Hmm, I wasn't sure, but financials weren't my strong point, so I moved on to something that was. 'And what about support?' I asked. 'If the customer has a problem with the application, who do they call? Us?'

Commercially, the customer would be ours and not the software vendor's, but they wouldn't get much in the way of application support if they phoned me. I'd just spent two whole days on that vendor's booth, and I still had no idea what their application did. The way Bruce talked about it, there was no limit to the number of different applications we could serve—but how would we support them?

'Just, don't worry about that stuff, that's not your job,' snapped Bruce, now sounding irritated. 'It's going to be FINE.'

Have you ever noticed that when someone says, 'It's going to be FINE,' it almost never is?

*

Back in London, Bruce hired more people. We had no customers or even serious interest yet, but he was gambling on a 'build it and they will come' strategy. He hired three young IT guys and rented a secure rack in a shared server room facility, where he wanted us to build servers and install applications, ready to go in case he convinced someone to do a trial.

Bruce tasked me with ordering the servers for the new guys to build, but when I brought him the invoice, he was outraged.

'What's this? I should have ninety days to pay for kit—I always got that when I was Head of IT at the newspaper! Go back and do it properly,' he snapped.

I asked the hardware vendor for the servers on three month's credit, but the guy said, 'No, sorry, we don't do that.' I tried another, and another, and got the same answer from everyone. One explained, 'We don't give credit to startups anymore, we've been burnt before.' The dot-com crash was making everyone nervous, and with good reason.

Bruce was furious when I told him. 'Give me that, I'll phone them myself. God, why do I have to do everything around here!' He grabbed his phone handset and stabbed in the number on the first vendor's invoice. I retreated to my desk, but it was a small office, and everyone could hear how his call went.

'So, I expect ninety days to ... what's that? ... Well is there anything that can be done about it? ... Right, OK, right, yes OK... Yes, we've got the invoice, I'll get my assistant onto it.'

So much for doing things right himself—he'd got the same answer I did, and he'd collapsed like a soggy meat pie. I had to take what satisfaction I could from that as I got no apology from Bruce.

This place was giving me some serious déjà vu. Bruce was clearly fretting about money if he was so upset about a couple of servers, and echoing what I'd seen at Ready2, his focus had shifted from building the business to searching for a new investor. He wasn't having any more luck.

'You've got to be Sir Rumpy Pumpy to get money out of vulture capitalists around here. If you're Australian, they don't want to know you!' he ranted.

The servers, paid for in full, arrived, and I worked with the new guys on getting them set up. And then, once that was done, there wasn't really anything else for us to do. I was reminded of another past job, of those weeks I'd sat around, bored and useless at Money Bank. I detested being in that situation again.

I tried to keep myself busy by writing documentation for our non-existent application service, and dreaming up plans for a proof of concept should Bruce ever convince a prospect to do one... but it was getting harder to maintain any enthusiasm. Without customers, without any real work, the whole thing had no more substance than a house of cards.

It seems counter intuitive, but underwork, especially for someone like me, is just as stressful as overwork—perhaps even more so, as I get none of the achievement buzz that keeps me going through tough times. And times were tough. With both Nathaniel and I working full-time, we were finding just how difficult it is when you have a baby and no local family support.

We were taking turns to drop Oscar at the childminder's house in the morning and pick him up in the evening. The mornings were horrible with getting a protesting one-year-old dressed and out the door. The afternoons were even worse with the rush to return to the minder's place on time. Too often I got stuck in a stalled Tube train, deep beneath the ground, a sick feeling in my stomach as my watch ticked past 6 pm, knowing the minder would be annoyed that I was late again.

In the evenings Oscar, well fed throughout the day, now refused to eat the food I prepared for him, and I felt like the worst, most useless mother ever. Weekends were a chore with the shopping and laundry. Without a washing machine we had to lug bags of clothes, towels, and bedding forward and back from the laundrette. I only stopped running when Oscar and I came down with yet another virus. I started to wonder what the hell I was putting myself through all this for.

One morning, about four months into the NewzServe job, I arrived at the office to be greeted by Bruce screaming that I was destroying his business. I was shocked, unable to understand what he thought I'd done through all his yelling and swearing. Eventually his assistant hissed 'Email isn't working,' and I thought, 'Seriously, is that all?' I rebooted the server and it started working immediately, but I was left shaky and upset by the ferocity of Bruce's attack. Again, there was no apology from him.

I'd thought Ready2 had been bad, but NewzServe had quickly become worse. Perhaps if I'd been doing anything that felt remotely constructive, the family time I was sacrificing might have seemed worthwhile—but instead my entire salary went on paying for a childminder to see more of my son than I did, and now I was getting yelled at to boot.

So demoralised was I that, for the first time since the programming sweatshop a decade earlier, I seriously considered giving up my computing career. I read that the UK government was so desperate for maths and science teachers they were paying people like me to do a Diploma of Education, with a guaranteed job at the end. I knew from experience that teaching wasn't something I felt passionate about, but I figured at least the hours would be more parent friendly. It was tempting.

The end of the year was approaching, and I didn't want to make any big decisions until January. Bruce had, some months earlier, booked a company Christmas party at a swanky London hotel. I knew this was a huge expense and had suggested that going to the pub for a pint would be just as festive, but Bruce would not back down. He was so wrapped up in his notions of success that there was no arguing with him.

On the day of the party, I felt such dread that I didn't want to go—but Nathaniel was coming too, and the babysitter was booked, so I dressed up nice and steeled myself for an evening of pretending that everything was FINE.

Once there, Nathaniel and I were shown to a grandly wood-panelled room. We were offered welcome drinks, followed by a sit-down, three-course meal. Everyone else seemed to be enjoying themselves, but I couldn't relax—it was all fake, and I couldn't bear it. To this day, the particular image I recall from that night is of Bruce getting up to waltz with his beautiful and much younger girlfriend, grinning confidently, his demeanour projecting, 'Look at me! I'm a success!' I still remember the leaden feeling in the pit of my stomach as I watched him.

Early in January, not long after we'd all returned to the office from the Christmas break, Bruce fired me with immediate effect. I wasn't surprised to be laid off, but I was shocked when he said there was no payout.

'Ready2 gave us all a month's pay when they laid us off,' I told him.

'I don't need to do that,' he said. 'If you're a small company and you're cutting staff for financial reasons, you don't need to give any notice or compensation.'

It was disorienting to suddenly see Trinny and Susannah in a far more positive light. I'd assumed they'd paid us a month's salary because they had to, but instead, they'd done it because it was the right thing to do.

'Anyway,' I said, 'I'm thinking of becoming a maths teacher.'

'Yeah, you should,' retorted Bruce. 'You'd do better at that.'

*

I had been in London just under a year and in that time I had worked for two failing startups, and I had lost two jobs.

For years afterwards, the story I told about my dot-com year was that Ready2 was a terrible idea with no possible way of turning a profit, while NewzServe was a good idea badly timed. Now, with the clarity of a lot of hindsight, I see little to differentiate them.

Neither company owned intellectual property that could be patented—the founders had ideas, but they depended entirely on the creations of others. When I stood at that booth in Amsterdam, it wasn't NewzServe I was demonstrating, it was Citrix. Likewise, Ready2 was nothing without the products of the high-street clothing stores they promoted for free.

The assumptions underpinning both businesses were sketchy at best. Trinny and Susannah gave away recommendations in the hope it would lead to a future payoff. Bruce expected to profit by undercutting the very software vendors he relied on. History has since shown that recommendations alone don't make money, and that software vendors will simply change their terms and conditions. Neither concept was ever going to work in the way they imagined.

Even had their ideas been realistic, these founders and their investors, like so many others during the dot-com boom, massively underestimated the myriad complexities involved in getting from a base functionality to a grand dream. They were people who had just been introduced to the concept of bricks and now planned to build the Taj Mahal. 'We'll get in a bunch of bricks and some builders and work out the rest as we go along. It's going to be awesome!'

During my dot-com year, I met a venture capitalist at a function and asked him why he invested in startups. 'I only need one out of every ten companies to get to the stock market to make a profit,' he replied. He wasn't particularly bothered about viable ideas or sustainable businesses. People like him walked away from the smoking ruins of the dot-com crash with tidy profits after cashing out at IPO time. Meanwhile, hundreds of thousands of people lost their jobs, and investors burned a combined total of five trillion dollars, much of it from personal investments, equity funds, and pension schemes.

I never invested, and my only financial loss was having to pay my childminder a full month's notice when I'd got none myself. But I'd lost in other ways. I was completely demoralised after a year of effort that had achieved no lasting or significant results. I'd sacrificed so much and it had all been for nothing.

I felt betrayed, let down by something I'd thought I could rely on when I moved to London. I had left so much behind in Sydney: my friends, my little brothers and sister, the familiar places where I'd been happy. My one consolation, my one link to my old life, was my career in IT. Now it had failed me. Right then, I wanted no more of it.

'Don't make any quick decisions,' counselled Nathaniel. 'Why don't you and Oscar go back to Australia for a visit. It will do you good.'

My heart lifted a little, he was right, that was exactly what I needed! I booked tickets, and a few weeks later, Oscar and I flew across the planet, towards the warmth of Australia and the people there whom I loved. It's such a cliché, but when the plane touched down at Sydney airport, and they played that damn song *I Still Call Australia Home*, I wept.

I spent five weeks back on home soil, reconnecting with the people and places I had missed so much. Nathaniel had been right, it did do me good. Five weeks was also long enough for me to really miss him, and for London to not look so bad after all, because he was there.

I thought a lot about what I could do better after I returned to London, particularly with work. I'll get to a major decision that I made in the next chapter, but I'll mention one here that specifically relates to Ready2 and NewzServe. Never again would I accept a job with some pie-in-the-sky, zero-revenue startup. I had made that mistake twice; there would not be a third.

16. How to be realistic

BEING BACK ON HOME SOIL WAS GROUNDING AFTER the unrealities of my dot-com year. Disappointments no longer seemed so bad while Oscar and I paddled at the beach and built sandcastles. Any idea of running away to join the teaching profession quickly dissipated in the Sydney sunshine.

One thing that didn't disappear was my desire to work. I was enjoying this relaxed time with my boy, but for the longer term, stay-at-home parenting would not be good for either of us. I could see myself, trapped in a small flat with a young child, increasingly frustrated and resentful. Without fully understanding why, I knew my mental well-being depended on having a job.

Back then, I hadn't yet been diagnosed with the Complex PTSD that my childhood left me with, so I hadn't attached that condition to the way my brain sometimes spiralled into a mess of fears, recriminations, hurts, and endless arguing of my case to a non-existent jury. I believe the official term for this is 'rumination', but that's far too gentle a word. From the way the thoughts toss around and around and around up there, my term for it is 'washing machine brain'.

Over the years I had discovered a few ways to subdue the spin cycle, at least for a little while. Wine, as long as I didn't overdo it, live music that I could get lost in, and properly absorbing work.

IT, despite all its frustrations, had given me the latter. There were always problems to solve, and there was always satisfaction to be found in doing so. My overactive brain found peace in the flow state of deep concentration. Given the space and quiet to focus, the disparate parts of a problem assembled themselves in my mind until I could almost *see* the connections between them. I needed that.

But something was going to have to change. Rushing to and from the childminder's had been hugely stressful the year before, to say nothing of what I had missed while at work. This included Oscar's first steps, caught by my father-in-law's camcorder for me to watch when I got home. My visiting in-laws had been so delighted that I'd had to pretend I wasn't horribly disappointed to have missed the real thing.

The obvious compromise was to work part-time, but I was reluctant. On some level I found the idea shameful, an unacceptable failure. I thought I had to be one of those Super Women that magazines pictured with a briefcase in one hand and a baby in the other, *having it all.* That shit gets in your head; it fills the spaces where your self-esteem should have been.

While I was in Australia, I talked this over with an old friend and she said, 'You know, you don't have to put how many hours you work on your CV.' Her comment stopped me in my tracks in such an Aha! moment that my shame about part-time slunk away in defeat.

After Oscar and I returned to London, I discussed my new plan with Nathaniel: I would keep working in IT, but only three days a week. We agreed that the trickiest part would be finding a three-day-a-week nursery place.

Only a couple of days later, it was almost as if something was telling us we were on the right track. Nathaniel came home with news that the Foreign Office was opening a staff nursery, right in the building he worked in. He'd already talked to someone, and they were happy to reserve a three-day place for Oscar. We'd lucked out with the childcare—now all I had to do was find a job.

I'd been searching for part-time listings, and I hadn't seen anything yet, but it was early days. For the next few weeks, I scanned the job sites and finally a part-time position appeared... but it was for a 'Junior Service Desk' role that I was ludicrously overqualified for, and which would not even come close to paying for Oscar's nursery. I did not apply.

My earlier confidence evaporated as the realities of part-time job hunting became clear. Had I got this completely wrong? Perhaps the only way to get a decent part-time job was to already have a secure full-time one to cut down. If that was the case, I was screwed. 'If we'd stayed in Sydney,' I grumbled to myself, 'Lew would have let me work part-time for sure.'

It never occurred to me to apply for full-time jobs and then negotiate. Recently, at a Women in Tech event I was attending, a speaker said that during her child rearing years, she had successfully negotiated several full-time positions into part-time ones. I was deeply impressed, both by the idea and her evident confidence in her own worth. Back in 2001, if I'd even thought of that, I doubt I would have had the nerve.

I had been back in London about a month when I invited my old Ready2 buddy Irini and her family over for Sunday lunch. I wanted to hear how she was doing at IBM, and to be honest, I also wanted some sympathy.

I was telling her about my fruitless job hunt when her husband, a university lecturer, stopped me with a suggestion. 'You should come and talk to our Chief Information Officer at London Business School,' he said. 'I can introduce you if you like?'

Yes I did like, very much, thank you. Irini's husband followed through, and an invite from the CIO arrived soon after. I could barely believe that I had, completely by chance, 'networked' my way to an introduction in a town where I knew so few people! You just never know what a friendly conversation might lead to.

I wasn't sure what to expect at my meeting with the CIO. Would he be genuinely looking for a new IT technician, or would this just be a casual chat, done as a favour to Irini's husband? So I was reassured when the CIO went straight into interview mode, asking about my experience and expectations.

I told him about the jobs I'd had and why I now wanted to work part-time. He then explained his plans to centralise and mature the university's IT and data services. I liked how he seemed to understand the current problems and complexities, while also having a clear vision for the future. I was interested to hear that, as part of this service consolidation, he had brought the Library under his IT banner—because what are books if not our oldest type of information technology?

At the end of our conversation, the CIO said, 'You have a very impressive CV, and we normally couldn't afford someone with your level of experience, but as you only want to work part-time, I think we can figure something out.'

On the spot, we agreed on three-fifths of what the dot-coms had paid me for three-fifths of the hours. And so began my six years of working for London Business School.

*

The best way to approach London Business School is through Regents Park. This way, your first impression is of the elegant Sussex Place, the university's main building. Constructed in the early 1800s and designed by the same architect who designed Buckingham Palace, Sussex Place is a palatial structure, painted entirely white from its stone base to the tips of the metal spires that stretch upwards from its domed, rooftop cupolas. As you walk up the gentle slope of its manicured front lawn, the broad sweep of the building becomes apparent, with its colonnades marching in either direction towards far wings that curve gently inwards around stands of shady trees. The sight is so evocative of Regency period dramas that you half expect ladies in floaty white dresses, parasols in hand, to emerge from the tall doorways and proceed to take the air.

Sussex Place looked like a fantasy, but I work in IT, and my day-to-day reality was going to be a nondescript sort of office building, off the main campus along a street of grey apartment blocks. As usual the IT Department was tucked away somewhere out of sight, in the less desirable real estate, while still close enough to be readily summoned should a printer be jammed or an application refuse to work.

On my first day, I felt a few flutters of nerves as I walked along the street to this building. I'd had many first days in my life, but it was still scary to walk into a place filled with people who already knew each other. I reminded myself to be positive: I'd had a hard time in London so far, and this job might just be the new beginning I needed.

During my dot-com year I had often worried about my career losing momentum. After the commercial projects of AusCM, my internal support and administration role at Ready2 had felt like a backwards step. I'd had high hopes that NewzServe would propel me into new technological and professional territory, but no such luck.

Now here I was, back-sliding even further, all the way to a university job—but I wasn't worried about momentum anymore. Instead, I felt proud of myself for making such a realistic decision. Consultancy was still out there, and I felt confident that I'd get back to it one day.

*

The CIO had assigned me to the Infrastructure Team which looked after the servers, networking equipment, and directory systems. I had been pleasantly surprised to learn that the team was headed by a woman, and my first stop that morning was to meet her.

One look at Laura's short, practical haircut and sensible shoes, and I knew we were going to get on. She further bolstered this impression with the best welcome I could have asked for: she already had jobs waiting for me. Because if there was one thing that scared me more than walking into an established team, it was sitting around awkwardly with no clue what I was supposed to be doing.

The first task Laura gave me was to set up a monitoring system to watch their computing infrastructure and alert on any failures. Laura explained that they'd purchased the monitoring application a while back, but no one had the time to set it up properly.

I'd never set up a monitoring system before, but I was up for it. 'That sounds like a good project for me, it will get me across everything,' I said to Laura.

She nodded, and then in the efficient and matter-of-fact way that I would come to really appreciate about her, she moved on to my next assignment.

'I also want you to find and assess any Microsoft servers we have,' she said. 'I only moved into this position a couple of weeks ago when the previous Infrastructure Manager left, and he had a bit of a problem with Microsoft. He was okay with having their products on the desktop, but he refused to accept we needed Microsoft servers. He always said we only had NetWare and Unix even though that's completely untrue. We have Microsoft servers, I just don't know how many or what sort of state they're in.'

The previous manager's anti-Microsoft stance was one I'd come across and even, at one time, shared myself. I'd started out in an IT world where Microsoft had cornered the PC desktop, while the server operating systems were Unix or NetWare. When Microsoft encroached on the server space I, like many others, said, 'Why on earth would I want to install *Windows* on a *server*?' But that had been years before, and things had changed a lot since then. I'm a realistic person, and I didn't see any point in sticking a stake in the ground while the market moved on. It had been in my best interests to move with it.

It sounded like the previous manager thought he could hold back the tide, and if so, he'd been kidding himself. 'Thank goodness he's gone,' I thought, 'I would not have liked having a fantasist as a boss.' Laura, in the fifteen minutes I'd known her, came across as the complete opposite: someone with her sensibly shod feet firmly on the ground.

Later, I met the rest of the team: five guys, mostly around my age of early thirties, with one junior in his twenties. The technical team lead introduced himself as Tomo. He would turn out to be a thoroughly lovely, gentle-natured guy with extensive technical skills and a remarkable ability to summon even the most obscure details about the university's IT systems to mind. He was held in high regard throughout the IT Department, and it was well deserved.

The others all introduced themselves, though most didn't say what they did. Only the junior, Duncan, cheerfully summarised his responsibilities as 'Backups and anything else no one wants to do.'

My first impression of my new colleagues was that most were the sort of men I was used to working with in IT—and by that, without any wish to offend anyone, I mean the type you could never imagine being captain of the football team in high school. Or playing football at all for that matter. They were the boys who'd grown up in the 'computer nerd' heyday of the 1980s and '90s. Back then, if you were a skinny and/or uncoordinated boy who was also decent at maths, it was generally assumed you'd have a knack with computers. This propelled many such boys into IT, while at the same time dissuading those who did not match the stereotype. Like girls.

Now, the IT sector was changing as the number of IT jobs exploded—a metric I'd been tracking through the responses I got when I told people what I did for a living.

In the early '90s it was, 'IT? What's that?'

Then there was a period of, 'What a coincidence, my brother-in-law works in IT! Do you know him?'

By 2001, when I started at London Business School, it was, 'Hmm, I hear there's good jobs in IT.'

(And now, in 2025, it's 'Yawn, yeah you and half the people I know.')

History has repeatedly shown that where there's money to be made, men flock in. In notable cases, including the early days of mainframe programming, they displaced a largely female workforce. The PC-centric IT profession that I had made my own had always been male-dominated, but by 2001, the expanding workforce was shifting the demographics. Not to more women though, just different types of men.

One of the guys in the Infrastructure Team stood out from the others. He was tall, fit, and definitely looked like he'd played football. He was even wearing his favourite footie team's shirt, and over the following weeks, I noticed that much of his office chit-chat revolved around sport. I remember thinking to myself then, 'When did IT stop being nerdy and start becoming blokey?'

I had my first conversation with this Footie Guy on my second or third day. I'd been looking over the Microsoft servers, as Laura had asked me to do, surprised to find that some of these 'non-existent' servers hosted functions as critical as the main School website. I had questions and had been told that Footie Guy was the person to see, so I went over to his desk and asked if he had a moment.

'I've been looking over the Microsoft servers,' I said, 'and I was wondering, what's your process when you set up a new one?'

'I just install the Windows Server operating system from the CD, get it connected to the network, and then I download and install the latest patches,' he said.

'Right, so, what about ongoing patching? Most of them are out of date.'

'No one ever said I had to keep patching them, only install them,' he replied, curtly.

'Well you know they need regular patching...' I started, and perhaps I was coming across as critical, because his arms were now crossed and he was scowling at me. I found him intimidating, but I had a job to do, so I pressed on.

'Also, I've noticed that most of the servers aren't licensed...'

Footie Guy cut me off. 'We have a site licence for Windows Server, that's what the old manager said. Are we finished here?'

Before I could ask why someone with a 'no Microsoft server' policy would have shelled out for a site licence, if there even was such a thing, Footie Guy had got up from his desk and stomped off.

I was a little taken aback by his hostility, but it wasn't the first time I'd had an unfriendly encounter with a man in IT. In fact, it was so common for newly-met techies to exhibit some degree of suspicion that I considered it normal.

This went all the way back to Geoff at the School of Electrical Engineering in Sydney. He was friendly and he taught me everything I needed to know, but there had been a discernible reserve in him, verging on irritation when I relied on him too heavily. Then, after I'd worked with him for several months, he took a long vacation, his first in many years. I was scared when he left, but over the weeks he was away, I had no choice but to figure things out for myself. After six weeks, Geoff returned to find that nothing was on fire, and that I was confidently running the show. It was then I felt his attitude shift: I had passed a test, and he always treated me as an equal and trusted colleague after that.

In the following years, as I moved through various jobs and projects, I noticed the same pattern time and time again: techies were initially reserved and suspicious until I passed some kind of unspoken test. The height of the bar varied greatly from one to another, and while I don't believe they were consciously testing me, it was always obvious when I'd passed. They relaxed their defences, started treating me as an equal, and I never had a problem with them again. So common was this phenomenon that I even had my own name for it: I called it the 'nerd handshake'.

Footie Guy might not have looked like the nerds I was accustomed to, but I figured his defensiveness was that same old handshake again. I trusted that, in time, I'd pass whatever his tests were, and then everything between us would be fine.

*

In my second three-day week, I went to see Laura with my list of Microsoft servers.

'There is no site licence,' she confirmed. 'Leave that with me and I'll make sure we're covered. As for the patching, are you able to work on a Saturday night? We have a regular downtime window on the first weekend of every month, 6 pm Saturday to 6 am Sunday. If you're not able to come in yourself, I'll get someone else to do it.'

In Laura, I had lucked out with a boss who understood exactly what it was to juggle work and parenting. She had a ten-year-old son who she was raising alone, and she knew how tricky it could be to arrange out-of-hours work, especially compared to the rest of the team who were all single, childless men.

I could have taken the free pass she was offering, but I knew that would be a huge mistake. To earn the trust of my new colleagues, I had to show I was prepared to roll up my sleeves and do the same work as them.

'I can do it,' I told Laura, 'no problem.'

My first Maintenance Saturday rolled around a couple of weeks later. At 5 pm, I kissed Nathaniel and Oscar goodbye, and I went to work.

I met the other members of the Infrastructure Team in a student lounge area in the historic Sussex Place building. There was no one else around, and the campus was eerily quiet. We spent a few minutes briefing each other on our plans for the evening, and then Tomo and Footie Guy headed off to another building to upgrade network cabling, while the rest of us went to the server room.

At that time, the server room was at one end of Sussex Place. If you followed the ground-floor corridor along the west wing, past the lecture theatres and right to the end, you found a locked door. Behind that door was the large, windowless room that housed our servers.

The room wasn't particularly well organised. The newer servers had been installed in a couple of free-standing racks, but the many older ones were in an assortment of un-rackable chassis, and they sat on desks, on mismatched shelving, or directly on the dusty floor.

That first Maintenance Saturday, I started with a circuit of the room to locate the servers on my patching list. I identified them by the sticky labels which, for some historically whimsical reason, displayed the name of the star, galaxy, or constellation that each had been named after.

Server patching has always been a tiresome process, particularly with servers as unmaintained as these were. The patches had to be installed in the right order, with many reboots in between. Some failed because prerequisite bits of software were missing, and you had to hunt them down, being careful to get the exact right versions.

Those old servers could take many minutes to boot, so I left one rebooting while I started on a second. Maybe even a third if the first two were both rebooting and it was particularly slow—the real difficulty was remembering where I was up to when I came back to each one. It was at once tedious and taxing work, and I welcomed the interruption when, around 8 pm, Duncan announced that dinner had arrived.

We all downed tools and returned to the student lounge, where takeaway Chinese food had appeared.

'This is nice,' I commented to Tomo.

'Yes,' he said, 'we always eat together on Maintenance Saturdays. It's good for the team.'

We tucked into the food while exchanging chit chat and a few laughs, though I still felt too much like an intruder to contribute much to the conversation. Dinner was quickly dispatched, the containers cleared away, and the break was over. Everyone was keen to crack on because the sooner we finished our jobs, the sooner we'd be able to go home.

Next, I started on a particularly decrepit looking old server. There were so many patches to install that it needed reboot after reboot, but that was too much for it. After the fourth reboot, it would no longer start.

I felt sick, the Chinese food curdling in my now anxiously twisting stomach. The spin cycle started up in my brain: I'd been arrogant, trying too hard to prove I could do the work, and now look what a mess I'd made of it. Oh god I was such an idiot and now they'd all know... etc. etc., rinse and repeat.

While that was all going on in my head, I was unplugging and replugging the cables at the back of the server and, after opening the chassis, the ones inside. But nothing seemed to have come loose. Whatever had happened, the computer was now dead.

There was nothing for it, I had to overcome my dread of asking for help. I took a steadying breath and then I called over to Steve, who was working in another corner of the server room. I didn't yet know him well enough to predict how he'd react.

'That old thing,' he shrugged, completely unconcerned. 'It was gonna go at any minute.'

He called Duncan over and asked him to fetch a spare computer. Duncan returned ten minutes later with a desktop PC that would apparently do. While I started installing Windows Server on it, Duncan tracked down the latest backup tape and got it mounted.

It took me the best part of an hour to get the server set up to the point where we could connect it to the backup system. I had no idea what the server even did, but fortunately Steve knew, and whatever it was, it was simple enough to get working once the data was restored. Even so, rebuilding that server from scratch was hours of unplanned work, and the three of us were there until 3 am getting it done.

It was an intense way to bond with my new colleagues, and not what I would have preferred, but it was effective. Back in the office on Monday, Steve and Duncan ribbed me in a friendly way about being a server killer, and I knew that meant they'd accepted me. There was still some reserve from the others, but I was certain they'd come around in time.

I felt great about working for London Business School. I had plenty to do, and it was practical work which delivered essential services to the university's staff and students. Far from those dot-com fantasies, this was real work with real outcomes that I could feel really proud of. And that, I realised, was all I'd ever wanted in a job.

17. How to be a pioneer

SILICON VALLEY ABOUNDS WITH MYTHS, INCLUDING that old one that technologists are far too rational for mythology. Silicon Valley wouldn't exist without its most foundational myth: that of the far-sighted pioneer, striding purposefully into uncharted technological territory, driven by a vision that we mere mortals are too stupid to understand. It's why they so love the quotation, falsely attributed to car maker Henry Ford, that 'if I'd asked my customers what they wanted they would have said faster horses'. It bolsters the myth by suggesting that regular people have no idea what they need, that only visionary technologists have that foresight.

While IT depends on the products of Silicon Valley and the like, we're a long way from that world. Our primary job is to maintain a computing environment that is stable and provides our users with the functionality they need, and this goal is rarely compatible with the leading edge of technology. Or, as we usually term it, with a wariness born out of experience, the 'bleeding edge'.

When I started working at London Business School in 2001, I was fresh off my dot-com year, and the bleeding edge held no allure for me. I was firmly in the 'don't be first' camp and yet, only a year later, I became a pioneer in something very new, something only made possible by technology.

It all started with a major Infrastructure Team project. I first heard about it a few months into the job when Tomo walked into our office and announced 'It's happening'. Based on the head shakes and resigned sighs from my colleagues, this was not welcome news.

'What's happening?' I asked.

'We have to move out of our server room,' Tomo explained. 'It's taking up space in Sussex Place and management want it for another lecture theatre. They've finally agreed to my budget for the new server room.'

'But what... why... how long?' I stuttered.

'It's OK,' said Tomo, looking amused at my dismayed expression. 'We're going to fit out a room in the basement of this building. We'll finally have proper security, aircon, redundant power, all that stuff. And they've just agreed extra funds to rebuild the old servers so we can fit them into the smaller room—it's a good outcome.'

It is not often in IT that we get to start over and do things right. Now that Tomo had explained the plan, my brain was whirring. I could rebuild all those haphazard old Windows servers with a nice, consistent, standard build, effective monitoring, and proper documentation. Lovely...

However, I was still worried about one thing: 'How long do we have to do all this?'

'Don't worry,' said Tomo, 'I made it clear that the server rebuilds are going to take time if they don't want a whole lot of unplanned outages. They're going to construct the new lecture theatre during the 2002 summer, so that gives us almost a year to get everything moved.'

OK! This was do-able! My first reaction on hearing Tomo's news had been concern, but now I was keen to get started.

Over the next months, while Tomo focussed on the fit out of the new room, the rest of us split the forty-odd rebuilds among us. Each of the guys volunteered for the servers they knew best, and I took whatever was left: fifteen or so, built years before by who-knows-who. There was no documentation and understanding each was an exercise in forensics, but it was just the sort of work I liked: getting down into the nitty gritty, tackling problems, finding solutions, and leaving things in a much better state than I'd found them.

Several months later, the new room was ready for business. The first servers to go in were our newer ones that were already in suitably slimline chassis. The guys simply unplugged them and walked them through the campus, across a main road, along the street to our building, and down the stairs to their new basement home.

Once moved, we got to enjoy a great feature of our new server room. Each rack had a KVM device (keyboard, video, mouse) which replaced the bulky per-server peripherals we'd had before. Now, we simply opened a menu and clicked through the list of computers in the rack to switch control to the one we wanted. And even better, the KVMs were on the network, and accessible remotely.

This was a game changer for us: for the first time, we could access a server from our desks rather than going and standing in front of it. It was a great incentive to get the remaining servers rebuilt and into the new room as soon as possible, and in the end, we had everything done well before the deadline.

Unexpectedly, this turned out to be a particular bonus for me.

*

After starting at London Business School, my home life settled into its new routine. From Monday to Wednesday, I went to work while Nathaniel took Oscar with him on the Tube into town, leaving him in the capable hands of the nursery staff for the day. On Thursdays and Fridays, I took Oscar to the playground or the library, and I got the shopping and laundry out of the way so that we could enjoy the weekend together.

Now that I had the time for it, I appreciated how much London had to offer with its parks, its free-entry cultural institutions, and its cozy, family-friendly pubs. On weekends, Nathaniel and I might walk up to Hampstead Heath and take Oscar to one of the playgrounds there, followed by a proper English pub lunch. In the summer we'd catch the tube to Hammersmith and stroll along the river, reminiscing as we passed the rowing club where Nathaniel had held his thirtieth birthday party. As Oscar grew, he became very interested in dinosaurs, and a trip to the natural history museum was always a winner on rainy days.

We knew that at some point, Nathaniel's career would take us abroad. As a general rule, a new diplomat spent three years in London before being assigned an overseas post, but there was a way to do it sooner. After his first year, Nathaniel became eligible to apply for jobs on the 'trawl', a list of posts that had to be filled quickly. He spotted one he liked the look of in Belgium and, with my blessing, he applied. He had very little expectation of being offered such a good posting so soon in his career, but much to his surprise, the assignment was confirmed. With only a few months to plan, we would be moving to Brussels in July 2002.

I was excited about the prospect of living somewhere new, and I can't deny that Brussel's reputation for the best chocolate shops in the world added to the appeal—but I was sorry to be leaving London Business School after barely a year. It was going to be hard to hand in my resignation to Laura, whom I greatly respected, and to say goodbye to Tomo and the rest of the team.

'It's good timing really,' I said to Laura, after breaking the news. 'I want to have another child, so perhaps three years off work won't be a bad thing.'

It wasn't true, that 'no bad thing' comment. The mere thought of being without paid work still gave me a squirmy anxious feeling, and not just for the immediate term, but for my future prospects as well. Tech, with its ever-changing landscape, was not kind to people who tried to get back in after several years away—I'd seen it happen.

'I'll be sorry to see you go,' said Laura. 'You're reliable and you do good work. I'll miss that, and I'll miss you too.'

We chatted for a bit about my plans and then, a little sadly, I returned to my desk to put my resignation, with eight weeks' notice, into writing.

A week or two later, Laura called me into her office.

'I've been talking to Tomo,' she said, 'and he says it's possible for you to work from Brussels if you want to. We can set up VPN access, and you should be able to do everything you can from your desk here. I know you'll be busy if you have another baby, so I'd like you to consider another option. If you switch to being a contractor, you can work as much or as little as you like and just bill us by the hour.'

Remote work wasn't unheard of in 2002—laptop manufacturers had already been selling us on the 'Work from the beach' idea for a few years—but it was limited. I might have occasionally worked on a document at home, but that was about all I could do. I'd never known anyone to work exclusively from home, but now we had our remote-control server room, it might just be possible.

'Thank you,' I replied, 'that sounds perfect!'

*

In July, we packed up our tiny London flat and moved to a much nicer apartment in a French-speaking suburb of Brussels. I enrolled Oscar, who had just turned three, in half-days at a local maternelle (pre-school). Then I set up my desktop computer and my ADSL internet connection in a corner of our living room, and while Oscar was at school, I worked.

From a technical point of view, Tomo had been right, and I could do much the same work from Brussels as I could in London. But there was a problem I hadn't anticipated. I had some ongoing responsibilities, but they only took a few hours a week and most of my work came from ad hoc requests, escalated support issues, and short-term projects. But now I was out of sight I was also out of mind, and the requests dried up.

Once a month, I caught the Eurostar train back to London and spent the day in the office. When people saw me, they said, 'Oh! Carol! There you are! Perhaps you can do this thing for me?' I realised that it wasn't just me who had to adapt to remote working—everyone else had to adapt to having an invisible colleague.

I started sending out emails ahead of my office days, asking if anyone wanted chocolates. People put in their orders, and my London mornings started with a walk around, collecting the cash and delivering boxes of Belgian deliciousness. It was a great opportunity for a chat, and work requests followed. I always went back to Brussels with a list of tasks to keep me going for the next few weeks.

Towards the end of 2002, my day back in London coincided with a celebratory dinner for the whole IT Department, so I decided to stay overnight. It was nice to chat with colleagues in a relaxed setting after seeing so little of them for months. As the wine and beer flowed, everyone became more animated, though no one noticed I was sticking to water.

After the plates had been cleared away the CIO made a speech. He thanked us for our hard work that year, and he asked that people stand as he mentioned major projects. I stood up with the rest of the Infrastructure Team to take our bows for the server room move.

When he ran out of projects, people called out other reasons for colleagues to stand up. 'Oldest person!' someone shouted, and a greying colleague reluctantly identified themself, getting a big cheer in return. Of course, the next one was 'youngest person!', with much hilarity as various people, egged on by colleagues, stood up and announced their age, only to find that someone else was younger.

We'd got to a junior in the Desktop Team, who was apparently the baby of the IT Department, when someone suggested that a colleague's newborn should win. Well, if they were going that way... I jumped up and tugged back my loose shirt so that everyone could see the bulge in my belly, which was roundly declared the winner.

That bump was our precious daughter, a little sister for Oscar, born in Brussels in 2003. We named her Pascale.

After she was born, I took a few months off. I had a much easier time, both physically and emotionally, than I had after Oscar's birth, and before long, my brain was itching to get back into something technical.

Day trips to London were off the cards while I was still breast-feeding, but my prior efforts to train my colleagues had paid off: when I announced my return, work followed. Shortly after, Tomo routed my office extension to my home phone in Brussels, and once people could easily call me, my work picked up even more.

I didn't work full-time, not even close. I usually billed around fifteen hours a week, and that was perfect. I got to keep my hand in, while still having plenty of time for my family. It had taken a year to settle into it, but remote working now felt like a genuinely viable alternative.

MONDAY 15TH SEPTEMBER 2003, BRUSSELS

My job this afternoon is to build a server for a new library application. Back in London, Duncan has rack-mounted the server and left it with the setup CD-ROM inserted, all ready for me to build. I'm going to spend the next couple of hours doing that, from 320 km away, on the other side of the English Channel.

I'm still amazed that I can do this at all. Perhaps the response times are a little slower than from my desk in London, but it's barely noticeable. Connected to the School's VPN, there's little I can't do from here.

I fire up the KVM client and connect to my rack. I see my new server in the list, and I arrow-key down to select it. The screen refreshes and now I'm looking at its console, setup program loaded, waiting for me to step through formatting the hard drive, configuring the file system, and joining it to the network. Then I'll install the operating system, and after that I have a list of other tasks: patching, securing, installing agents and utilities, setting up standard file shares, and a base level of monitoring. It's a solid two to three hours of work and I'm hoping to get it all done this afternoon.

But it won't be without interruptions. Helga, my babysitter, arrived just after 1 p.m. and has been looking after six-month-old Pascale for the last hour. Her name may conjure the image of a tall, blonde, Germanic lass, but our Helga is actually a petite, dark-haired Chilean. She studies at college in the mornings and comes to look after my kids in the afternoon. Soon, she'll pop Pascale into her pushchair and head out to pick up Oscar from school, but first, she brings her to me. I pause what I'm doing and relocate to an armchair where I can comfortably breastfeed my baby. Helga perches on the couch nearby and we chat about how she's doing in school, and her plans for the future.

After Helga and Pascale have left, I get back to the server. I should have at least an hour, perhaps more if Pascale falls asleep in her pushchair, and if Oscar wants to stop at the playground on the way home. If I keep my focus, I'll be able to knock through all the tasks on my build list before the door flies open and Oscar rushes in, full of his day, happy to see me, wanting a cuddle and a snack, and the work part of my day will be over.

*

If there had been any resistance outside the Infrastructure Team to the idea of me working remotely, Laura dealt with it and I never heard about it. Only Footie Guy let me know with one of his snarky asides that he viewed this opportunity as blatantly preferential treatment—but he always deferred to Tomo, whose idea it had been in the first place, so he never came right out and protested it.

Anyway, it was beneficial for him too. Later, after Nathaniel's three year posting ended and we moved back to London, I returned to the School to find that everyone in the team now worked remotely from time to time. It had become normal to say you were working from home if you had an appointment, or you had to wait for a plumber, or you'd woken up feeling a bit rough. I am not taking credit for this, but I believe my pioneering example, my proving that remote could work, accelerated its adoption.

In the years following, remote working became increasingly common in IT, but it surprised me how little it took off elsewhere. It wasn't until 2020, eighteen years after I moved to Brussels with an office PC and some network cables, that 'telecommuting' finally hit the big time, all thanks to a little virus that sent white collar workers around the world scurrying for the safety of home.

During the Covid pandemic, as everyone got around video conferencing, team messaging, and virtual desktops, it amused me how many people seemed to think this was an entirely new and revolutionary thing. 'Remote working could actually be really good for people with caring responsibilities,' commentators were saying. Yeah, tell me about it, I'd been saying that for years.

I'd also been saying that flexibility was the perk that made IT a particularly good career choice for women. What other job, in 2002, would have allowed me to so neatly balance work and parenting? How else could I have kept up my skills while still being on-hand to care for my children? I felt luckier than ever to have found my way to IT. I hoped that many more pioneering women would follow, and that they would also discover the benefits that I had.

18. How to be a Woman in Technology

REMOTE WORKING FROM BRUSSELS HAD ALLAYED MY anxiety about a resumé gap, but it didn't take me long to replace it with another. I would be turning thirty-six around the time we moved back to London at the end of Nathaniel's posting, and a particular landmark birthday no longer felt so far off. I worried about my career prospects once I became *a woman over forty*.

I had long believed that I would 'age out' of technical work by then. I didn't think I'd suddenly stop being capable, but I worried that employers would think that of me.

This was not an idea I came to on my own. In my twenties, I regularly heard hands-on tech described as a 'young person's game'. Older colleagues would say to me, completely unprompted, 'You know, you can't do this type of work forever. You need to be thinking about what you'll do next.' And then they'd suggest management or training or sales. Anything, apparently, but technical work.

I didn't know if my male colleagues were the recipients of similar unsolicited advice, but either way, they seemed to have a point. In the nineties, most of the techies I came across were in their twenties and thirties, so perhaps there was an age limit. It's only looking back now that I see the obvious connection: we were young then because our profession itself was young.

Until my mid-thirties, I deflected suggestions that my gender might have an effect on my career—but I only had to turn on the TV, and see whose careers continued as their hair greyed, to know that ageism was compounded by sexism. If IT was indeed ageist, it would certainly be harder on me.

When I did come across older women in IT they were, like Laura, in non-technical positions. Was that the only way forward for me too?

One thing was certain: I would not be following Laura's footsteps into management. I had watched her in action and admired how she set clear goals, navigated disagreements, and developed each team member according to their unique preferences and temperament. I knew I wouldn't be like that. I could just see myself getting too lost in the details to give clear directions. Or not giving directions at all because I assumed everyone would figure out what needed doing and then go do it, as I always had. And then, when they failed to be Carol Clones, I'd get annoyed with them. No one needed that.

But there was another type of manager in IT.

'I'm thinking about retraining as a Project Manager after I get back to London,' I said to Laura on one of our regular phone calls.

'Why would you want to do that?' asked Laura.

'Well, it's just, I'm not getting any younger and I don't know if I can keep doing technical work,' I said.

'That would be a total waste,' she said. 'You're so good at what you do. You should keep doing it.'

Really? It was that simple? Just keep doing it?

Sometimes, the right words at the right time are all it takes for your house of cards to fall down. I realised that I had no proof of this 'aging out by forty' thing, so why not test it? After that chat with Laura, I stopped considering alternatives. When we moved back to London, I would keep doing what I did.

*

As our departure from Brussels drew nearer, Laura rang to talk to me about a new project.

'We're moving email to Microsoft Exchange,' she said, 'and when you get back to London, I want you and Steve to design and build the new email servers.'

This was big! For the last three years, I'd been doing mostly routine stuff—which had been fine, but once the novelty of working remotely wore off, not especially interesting. Now, with the prospect of a big project to sink my teeth into, I felt a tingle of the old excitement.

Laura went on to acknowledge that neither Steve nor I had any experience with Exchange.

'So that is why,' she said, 'I'm sending you both to the Microsoft TechEd conference in Amsterdam, so you can learn all about it. And also, it will do you good. Get your head back into work mode.'

I couldn't believe my luck, a new project *and* a conference! Conveniently, Nathaniel's parents would be visiting right when TechEd was happening, so I didn't even have to worry about the children. Things were falling into place, and I took that as confirmation that staying the course with hands-on tech was the right decision.

A couple of weeks before the conference, I went through the agenda, ticking sessions. Anything related to Exchange, of course. Otherwise, I was most interested in cross-systems integration.

A big challenge in IT is getting products from different vendors (or even the same vendor) to play nicely together. We have better interface standards these days, but back then it was like the early days of rail travel, when passengers had to switch trains in odd places because the track going forward belonged to a different company and was a completely different gauge. IT incompatibilities are just as much of a nuisance as those train changes must have been, and I was interested to hear about creative solutions.

After selecting all the Exchange and integration sessions, I scanned the other tracks to see if anything else popped out. What I spotted sent a thrill through my body: it was the 'Women in Technology Lunch'.

I had now, in 2005, been working in IT for fifteen years, and I still needed only one hand to count the women I'd met who, like me, did technical systems work. The thought of being in a whole room full of them was almost too amazing to be true!

In the weeks leading up to TechEd, I daydreamed about the Women in Technology Lunch. I imagined hanging out with my fellow techie women, how we'd swap war stories, how we'd impress each other with our savviness and our sassiness and become life-long friends. Perhaps I'd even meet an older woman, still killing it with her tech skills, and she'd take me under her wing and tell me how it was done!

I was a thirty-five-year-old mother of two, but inside, I was as excited as a thirteen-year-old looking forward to her first high school party.

*

I'd been to a couple of tech conferences before, including Australia's very first Microsoft TechEd in 1994, but I'd never seen anything like TechEd 2005 in Amsterdam. With 10,000 delegates, hundreds of exhibitors, and a vast array of sessions across multiple tracks, it took me most of the first day just to get my bearings.

I dived in, going to session after session, my head filling with ideas. During the Exchange talks, Steve and I sat together, scribbling notes and rough architecture diagrams. Afterwards, we'd ask the speakers for their opinions, and they were always happy to advise. Already, we had a good idea about what our design would be.

My most-anticipated event arrived on day two. I farewelled Steve, reminding him for the six hundredth time that I had other lunch plans that day. And then, feeling happy and special and perhaps a little important, I made my way to TechEd's most exclusive gathering.

On entering the room, I noticed that one end had been set up with a screen and a few rows of seats, while the other had a buffet and some tables and chairs. I headed straight for the buffet, filled a plate, and sat down at a table with a handful of other women.

I had barely exchanged names with them when we were interrupted by someone standing at the other end of the room. Someone with a microphone.

'Ahem, excuse me, everyone, can you all bring your plates and sit down over here,' she instructed, waving at the rows of chairs facing the screen. I was confused—I had been expecting a casual lunch and not another session.

Everyone obediently moved, so there didn't seem much I could do other than follow. While we took our seats, Microphone Lady clicked something and Helen Reddy's voice, singing 'I am Woman', blasted out over the speakers. 'Of all the obvious,' I muttered irritably to myself.

There was more obvious to come.

Once we were seated, Microphone Lady kicked off a PowerPoint slideshow which looked to have been assembled after five whole minutes of internet searching 'famous women'. She'd included Marie Curie and Amelia Earhart, but not a single woman with the slightest connection to computing. Having not yet heard of computing's greatest women, I would have been thrilled had this slideshow introduced me to Grace Hopper—legendary programmer of one of the world's first computers, co-creator of COBOL and tireless advocate for women in computing. I would have been intrigued had she included Hedy Lamarr—Golden Years of Hollywood bombshell and inventor of secure wireless communications. Even Carly Fiorina was missing, and at that time she was the CEO of computer hardware giant, and conference sponsor, Hewlett Packard!

It was too frustrating. All I'd wanted was to chat with other women, but instead, as Microphone Lady was pleased to inform us, we'd be listening to two speakers.

The first was a business grad who happened to work for Microsoft in an entirely non-technical capacity. Her presentation was on things her male boss had said to her that she thought were profound. She tried dressing it up for a nerd audience by referring to him as 'my personal Yoda', but referencing *Star Wars* in no way improved this business buzzword twaddle.

The second speaker was a man with the creepy-sounding job title of 'Microsoft Evangelist'. He told a rambling story, the point of which seemed to be that he'd once hired a woman and hadn't regretted it. It was offensively patronising.

After the Evangelist slouched off, Microphone Lady asked if there were any questions. Almost of its own volition, my hand shot up. She motioned for me to speak, and I wasted no time getting to the point.

'I have sat here and listened to a woman in *business*, and a *man*,' I said, 'but I was expecting to talk to other *women in technology*. This is, after all, TechEd—an educational conference for *technical* people.' I turned to the forty-odd women sitting in the rows of seats. 'Who else here does a technical job?' Hands went up. 'I wanted to talk to *them*, not listen to whatever that was!' All around me, heads were nodding in agreement.

'But, but we have Olga as well!' countered Microphone Lady. She ushered on, from the sidelines, a very reluctant looking woman, who took the mic and began reciting her CV. Olga did at least seem to be the real deal, with a background in development and project work and training, but that only made it worse. 'If they had Olga there all along,' I thought, 'why did they make us listen to those other two?' I waited until she had finished speaking, and then I pointedly got up and left.

I was furious—I had been robbed! I went to that lunch to talk to other women in technology, and I had been actively prevented from doing so.

I marched off to my next session, but I couldn't concentrate, I was too angry and disappointed. After ten minutes, I gave up and went to a feedback terminal, where I consoled myself by giving that Women in Technology Lunch the worst possible scores. The damn thing didn't offer zeros, so I settled for all ones, along with a scathing comment.

And then a funny thing happened. Throughout the rest of the conference, women grabbed me as I went past, saying, 'That was you at that lunch! That was great what you said!' Apparently, after I left, Microphone Lady finally relinquished control and allowed the attendees to talk to each other. There had been a good discussion, these women said. I was sorry to have missed it, but perhaps I had to walk out for it to have even happened.

*

On the final morning of the conference, as I was loitering near a coffee station, another woman approached me.

I was hungover that morning after the conference party the night before. This woman, in contrast, looked clear eyed and well-rested as she said, 'It's you! From the Lunch! That was great what you said!'

We exchanged names and then she asked, 'What are you doing now?'

'I'm not sure,' I said, between gulps of coffee, 'I didn't have anything marked for the first session today.'

'Why don't you come with me?' she suggested. 'I'm going to one about application security.'

Still feeling woozy and clutching yet another coffee, I trailed along with her to a big theatre. It was full of people who, like my new friend, looked far too perky after the festivities of the night before.

The session wasn't one I would normally have gone to. In IT, there had long been a divide between Development, or Dev, and Operations, or Ops. I was firmly in the Ops camp, and this session was a Dev one, but I figured I could just sit there and let it wash over me for the next forty minutes.

The speaker had only one point he wanted to get across, and he said it so many times, in so many different ways, that even I in my hungover state got it loud and clear. Essentially it was, 'For goodness sake people, stop developing as Administrator because you end up with applications that only work in Administrator mode!'

'Devs are so dumb,' I thought from my typical Ops viewpoint, 'they don't get security at all.'

In the Q&A someone asked, 'What if I have to do my development on a server through Remote Desktop? You have to be a full Administrator to use Remote Desktop, right?'

'Oh, yes,' said the speaker, looking a bit deflated, 'that is a problem. I don't really know what you can do in that situation, sorry.'

'Ha,' I murmured to my new friend, 'he doesn't even know, it's so easy. You just put them in the Remote Desktop Users group.'

She turned to me, looking far more animated than I, in my current condition, could imagine ever being again. 'What?' she said. 'You know the answer? You have to tell them!'

Before I could stop her, she'd grabbed my hand and was dragging me to the aisle and down the stairs. A microphone had been set up for questions, but she marched me straight past the queue and towards the front of the stage, where she gestured urgently to the speaker.

'She knows!' she said, pointing at me. 'About the Remote Desktop question, she knows the answer!'

He waved someone over and I was handed a mic. I turned to face the packed theatre and, trying to quell both the queasiness in my stomach and the crack in my voice, I said, 'The Remote Desktop thing, it's easy. There's a built-in group on every server called Remote Desktop Users, so just ask your sys admin to add your account to that. You don't need Administrator access.'

I wasn't expecting the round of applause. To me it was such a trivial thing, but that's IT for you, we can't know everything. The Devs didn't know simple Ops stuff, and I didn't have a clue what they did either.

And then, with that particular question set to rights, my new friend and I walked out of the theatre arm in arm, my hangover completely gone.

*

I don't remember that woman's name, but thanks to her, and to the other women who stopped me in the hallways, I did learn something about how to be a Woman in Technology at that conference. I learnt that there is power in speaking up, and even more in supporting and celebrating each other.

Her obvious pride in me and my little piece of pertinent knowledge was both gratifying and encouraging. Had she not pushed me forward, I would never have told a whole theatre full of Devs something they didn't know. I would not have walked away thinking, 'You know what? I *am* good at this, and I *can* keep doing it as I get older, just watch me!'

But even with those positive interactions, that Lunch left a bad taste. Years later it still infuriated me that its organisers thought we women would have so little to say to each other that it was preferable to fill the air with slapdash and ill-judged presentations. It would have been infinitely better had they done the absolute minimum: a free room, a few plates of sandwiches and leaving us to get on with it would have been perfect. Instead, we had to sit there and be told how to be Women in Technology by people who knew fuck all about it. The experience was so painful that it was many years before I could bring myself to go to any kind of Women in Tech event again.

When I originally wrote out the story of that Lunch, I posted it on Facebook, thinking it might give some of my tech friends a chuckle. I was surprised at how many responses I got from female friends in all areas of STEM.

'Groan, that was too familiar,' said one, and, 'I refuse to go to those things,' said another. All had their own experiences of patronising and irrelevant 'Women in ...' events. One said, 'The conference organisers always put the women's thing on at the same time as the science talks we're there for, and then wonder why nobody goes to it!'

Seven years after Amsterdam, I was invited to speak at the Women in Technology Lunch at Microsoft TechEd 2012 in Australia. Before accepting, I asked how much time attendees would have to talk to each other, and only agreed once they confirmed there would be plenty of time for chat before I took the mic.

Then I put together the sort of presentation I would have loved hearing back in 2005: one in which a woman *over forty* talks about her long-term career in tech.

There is something to that adage, 'You can't be what you can't see'—it's not a strict rule, but it's certainly *harder* to be what you can't see. I've rarely met women older than me in hands-on tech, and I don't wish that on the younger women in my profession. Visibility counts, and that's a big reason why I wrote this book. I'm an IT elder now, and the least I can do is be seen.

19. How to love technology

I COULD NOT HAVE MADE A LONG-TERM CAREER IN technology if I'd never fallen in love with it.

Because, and I hate to tell you this, working in IT can be total and utter bullshit. We're beset by bad software that is badly configured and badly maintained, compounded by idiotic decision-making, and corner-cutting that is so herculean it could be an Olympic sport. And don't get me started on the relentless pursuit of cost savings over quality which always seems to cost more in the long run. The sorry truth is, you can't work in IT without being frustrated, exasperated, infuriated, beaten down and, on occasion, completely broken-hearted.

While I can be as cynical about this as the next techie, underneath it all there's still love, and it's kept me going all these years.

To be clear, I don't love technology for its own sake. To attract my affections, tech must answer a genuine need, and do it in a way that is functional, reliable, and elegant. It must 'do what it says on the tin' in a coherent and useful way.

Take email for example. These days I'm sure you're as sick of all the garbage clogging up your inbox as I am, but in the early days email was a marvel.

I remember, like it was yesterday, the first time I saw a photo delivered by email. It was in 1991, the proto-internet days, and the photo was of a newborn baby. It had been sent to my colleague Geoff by the proud father, an old friend of his, from the other side of the planet.

Before then, the only way to send a picture was by post. You had to shoot a whole roll of film, get it developed, pay extra for additional prints, pop one into an envelope, locate your mate's home address so you could write it on the front of the envelope, and then take it to the post office, because who the hell knew how many stamps you'd need to get it to Australia.

Being able to send a photo by email was a game changer. No need for extra prints or stamps, you only had to know your mate's email address and the photo arrived within moments of you sending it.

Even so, it wasn't easy. Only a minuscule fraction of the world's population had an email address in those early days, and if you were one of them, you also needed access to a fancy scanning device to create an image file from your printed photo. The next problem was that the email could only contain text, and attachments did not yet exist. To solve this, some clever person had come up with the idea of encoding the image data into a long stream of letters, numbers, and symbols that could be pasted into an email.

If you knew where to look, there were free programs you could download that converted the image to text and then back again.

Geoff explained this encoding and decoding process to me as we watched the picture of the baby slowly appear, line by pixelated line, on his screen. It was mesmerising.

I marvelled at how this picture had travelled thousands of kilometres, through cabling and satellite and yet more cabling, to arrive in Sydney, Australia from San Francisco, USA. So far away, and yet it had been sent only minutes before. It felt like the planet had folded in on itself and now Sydney and San Francisco were near neighbours. The world was shrinking, in front of my very eyes, to one where a moment of such tender humanity could be so easily shared with loved ones far and wide. I could imagine only good things would come of this.

Days later I was still thinking about the photo, turning the memory over in my mind, examining it from all angles like a precious jewel. I remember sitting with a friend in the cool night air outside a pub, taking a brief moment of respite from the heat and noise of the rowdy gig going on inside, and I told her about it. Normally I didn't talk about work with my band scene friends, but this story was still so uppermost in my mind that it just came out.

My friend, like most people at that time, had never heard of email, but she immediately understood how big this was.

'So, let me get this straight,' she said, 'the photo came from the USA, and it took how long to arrive? Minutes? How is that even possible? How did the photo know where to go? This is absolutely amazing!'

She was right, it was amazing, and there was much more. In the early nineties, ground-breaking innovations seemed to rain down from the sky. Like in 1993, when the theoretical physics research institute CERN made the technology that powered the World Wide Web generally available, followed later that year with the release of Mosaic, the world's first free web browser. The possibilities of a networked world seemed endless then, even utopian. While we can have many discussions about how that all turned out, one thing holds true: the early 1990s was a fantastic time to fall in love with technology.

*

In mid-2005, soon after TechEd Amsterdam wrapped, Nathaniel and the kids and I moved back to London. After four years of part-time working, one in London and three remotely from Brussels, I was going to work full-time at London Business School.

There was something in particular I hoped to find in the office. It had been so long since I'd felt anything approaching those highs of my early days in IT, and while I appreciated this career that had kept me employed through three international moves and two babies, I had, somewhere along the way, mislaid my passion. I missed it like an old flame and I hoped that the Exchange project would reignite the spark.

I had told Laura that I'd be back in the office a week or so after our arrival in London, but that proved ridiculously optimistic. Nathaniel had to go straight to work, and it fell to me to find us somewhere to live, a school place for Oscar, and a nursery place for Pascale. I pushed back my start date and then pushed it back again. By the time I made it into the office, the Exchange project was well underway.

I was crushed to find I had missed the design phase. Steve had done an excellent job of selecting and sizing the servers and storage system, and had already ordered the hardware, but that didn't stop me acting out my disappointment by going through his design with a fine-toothed comb and questioning all his decisions.

Hoping to claw back some role for myself, I suggested I build the servers when they arrived, but Steve made it clear he'd be doing that. 'Don't you already have enough to do?' he said. But the work I had was the boring stuff I'd been doing in Brussels, and I hadn't come back full-time to keep on doing that. So I badgered Steve even more, trying to insert myself into work he already had well in hand, much to his irritation. Other members of the team commented that the pair of us were 'bickering like an old married couple'.

Eventually I realised that all I'd achieved was making a nuisance of myself. If I wanted to make a significant contribution, I was going to have to find a way to do so without stepping on any toes. I scrutinised the project plan, seeking something others had missed.

The plan was to move mailboxes in daily batches, spread out across many months. This staged approach is common with business-critical systems like email, as it's less risky than a 'big bang' migration, and there's more time for user hand-holding. But staging is a lot more complicated for IT because the old and new systems have to coexist, and that is rarely straightforward.

The first mailboxes we moved were our own—we always start with ourselves so we can iron out issues on a forgiving group of users. I watched as three colleagues tag-teamed the jobs of creating the new email account, changing settings on the old one, modifying address lists on both systems, updating mailing lists, correcting mail routing rules, and finally, moving the mailbox contents. As they went, they called out to each other, advising when something had been done on one system, so that the next step could be taken on another.

This process struck me as both inefficient and highly risky. While the migration progressed, we still had our regular work to do, and it was inevitable that attention to detail would waver under the onslaught of other demands. Steps would inadvertently be skipped or mistakes made, and this would not only waste everyone's time, but make IT look incompetent in a very visible project.

This was it, the thing that no one else had thought about! The only way we'd get through this extended migration without everyone losing their minds was by automating as much of it as possible.

My biggest hurdle was figuring out how to coordinate changes between systems that were like different countries with different languages and laws. I could use some basic scripting tools to automate a few steps, but they only worked in a single system. What I needed was something that could talk to everything in its native language.

My research brought me to Microsoft Identity Integration Server, or MIIS. This product was designed to manage user accounts, or digital 'identities', across disparate and incompatible systems. As I scanned the list of what it knew how to talk to, it appeared to be exactly what I needed. I downloaded the software and got straight into building my first MIIS server.

I quickly found that MIIS is not simple. It is essentially a platform on which you build your own solution, and out of the box, it doesn't do much. I figured out how to link together a person's accounts from different systems, and how to copy a few attributes from one to the other, but I needed it to do much more if it was going to be of any use on the Exchange project.

I studied the very limited documentation until it dawned on me: to make MIIS do anything interesting, I was going to have to *write code*.

This was not a welcome realisation. Despite having learnt, taught, and even briefly worked in programming, I hadn't written code in years. In IT's Dev/Ops divide, I had aligned myself firmly with the Operations camp, and meanwhile, development tools and methodologies had moved on a lot. I didn't even know where to start with writing code these days.

Luckily there were people nearby who did: the team of developers who maintained our inhouse applications. I asked if any of them might give me a hand, and one lovely chap agreed.

Over the next few weeks, I spent as much time with him as he could spare. He figured out how to write the very specific style of code that MIIS required, and I explained exactly what I needed it to do. Together, we got the solution to a point where it was ready to help with the mailbox migrations.

Now, instead of a tag-teamed list of manual tasks, all someone had to do was set a flag on the old email account and MIIS jumped in, creating the new account and doing most of the fiddly backend changes. And if a user's migration needed to be delayed for some reason, a simple reset of the flag told MIIS to reverse its changes.

I had significantly reduced the manual work in the migration while ensuring the steps were done in exactly the same way each time. I had also become completely smitten with MIIS. I couldn't get enough of watching it going about its business, executing to perfection the rules I had configured, my obedient little transformation engine.

*

Putting MIIS to work as an email migration helper was not a typical use. As an 'identity management' system, MIIS was designed to automatically create user accounts for people when they joined the organisation, update their access as they moved through different jobs, and then deactivate their accounts when they left.

I went to Laura with a proposal: 'MIIS is being very useful on the email project, but there's so much more we could do with it. For starters, there's all the different accounts that have to be created for students and then disabled when they graduate—I could completely automate that!'

Laura was, as ever, prepared to support a good idea. From that point on, building out the MIIS solution became my primary role.

MIIS took me back to the best parts of my early days in tech. Focused intently on my solution, I was in my happy place: working at my own pace, setting my own goals and then knocking them down. MIIS is a very difficult product to master, and I made many mistakes, but I learnt fast and I was hungry for more.

By the second half of 2006, a year after returning to the office, I had automated all staff and student account creation and termination in our major directories and applications. In the process, I had freed colleagues from repetitive manual work, in particular a young woman whose job it had been to bulk create accounts for incoming students—an error-prone and often late-night operation. She was delighted not to have that task anymore, and suggested other jobs that MIIS might do. Later on, after I left London Business School, she took over responsibility for the solution.

MIIS had reignited my passion for tech, and it had also given me an idea about my future career. I was happy with my decision to stay 'on the tools', but if TechEd Amsterdam had shown me anything, it was that there were 10,000 men out there, most of them younger than me, all potentially vying for the same positions. If I wanted to stay relevant beyond forty, I needed a way to stand out from the crowd. My best bet, I decided, was to become a specialist—a MIIS specialist.

Nathaniel and I were expecting to be in London for another year or two before our next overseas posting and that should, I figured, give me plenty of time to plan my transition from IT generalist to specialist.

And then everything changed again.

20. How to roll with the punches

THE LAST MONTHS OF 2006 UNRAVELLED IN A series of gut punches.

The first came when Laura informed us that she would be leaving in a couple of weeks. She had been offered a job in a larger university with more opportunities for growth. As it later turned out, this was a great move for her, and she would go on to become the CIO.

I was sad to see her go. It had been almost six years since I'd joined her team, and what had started as a collegiate and mutually respectful working relationship had grown into friendship.

There was, however, a silver lining to her departure. I would need to leave to pursue my specialisation goals, but my loyalty to Laura had been holding me back. Well, not anymore. On my lunch break that very day, I rang the only specialist MIIS company I knew of and asked if they needed another consultant.

Conversations with that company had only just begun when the next surprise came. Nathaniel was offered a four-year posting to Geneva, starting midway through the following year, 2007. We would be moving in nine months, and I didn't like the idea of taking a new job knowing I'd only be there for a short time, so I put my specialisation plans on hold.

And then, shortly after Laura left, came the worst shock of my working life. It was delivered by Footie Guy.

My relationship with him had never settled into the comfortable familiarity I enjoyed with the rest of the Infrastructure Team. I tried to be friendly, grabbing at any scrap of common ground in my efforts to build an accord with him, and sometimes it seemed to work. We'd have a good day, tackling a job together, kicking goals, and after that I'd think 'Phew, he's OK with me now.' But it never lasted. Before long he'd be back to scowling and making nasty asides, and I took the hint and kept out of his way as much as possible. I knew he was talking trash about me behind my back, but I shrugged it off. 'Let people make up their own minds about me,' I told myself.

Until she left, I had no idea just how much Laura's presence had been keeping him in check. Once she was gone, there was nothing holding him back. Only three days after her departure, Footie Guy let loose on me in the most aggressive way I have ever experienced in a place of work.

I can't remember what triggered him—all I remember is that I was on the phone with my good mate on the Service Desk, helping him out with something, when suddenly Footie Guy was standing over me in a towering rage, screaming obscenities and looking like he was seconds away from punching my head clean off.

I was terrified.

Footie Guy was a tall and muscular man who made no secret of his disreputable past as a football hooligan. I knew he was capable of violence because he had told me so himself. And right then, he looked unhinged enough for it.

Instinct took over. I dropped the phone handset, grabbed my bag and fled. I ran out of the office, and out of the building, and down the street. I didn't stop running until I was well on my way home.

A few minutes later my mobile rang—it was my Service Desk buddy. He'd heard the whole thing through the phone and wanted to check I was ok. He listened patiently while I cried and hyperventilated, and then he said soothingly, 'Don't worry about him. He's a dickhead but he won't do anything.' I did not feel reassured.

That evening, I told Nathaniel what had happened. He looked horrified.

'You don't know what this guy's capable of,' he cautioned. 'I think you should call the police.'

I considered it for a moment, and then I rejected the idea.

'He might have looked like he wanted to punch me, but he didn't, so I don't think they'd care,' I said.

Nathaniel tried a few more times, but I wouldn't budge. And I still believe what I said then, that without physical wounds the over-stretched Metropolitan Police would have brushed me off. Or worse still, they'd come to the office asking questions and that would only inflame the situation. I'd end up the bad guy, of that I had no doubt.

Apart from his thing with me, Footie Guy was well-liked by the rest of the team. A number of them had witnessed his outburst, but they didn't check in on me. I knew they'd all be minimising what had happened. They'd be giving Footie

Guy the benefit of the doubt and framing the situation as being of equal fault. Even though I'd been the one terrified out of my wits.

So, I would not call the police, but I wasn't going to let it slide either. I was going to make a complaint about Footie Guy to our manager.

I'd barely met the new guy, and I felt bad about dumping this crap on him, particularly as he was a temp, hired on a short-term basis until a permanent replacement for Laura could be found. But he was the manager, and dealing with staff conflict was part of that job.

The next morning, my guts still twisting with the anxiety that had kept me awake most of the night, I phoned the temporary manager.

'You've heard about what happened yesterday?' I asked.

'I heard there was an argument,' he said.

'That wasn't an argument, it was an attack,' I stated, but already there was an embarrassing quaver in my voice. 'I'm too upset about it to come in today,' I sniffled. 'I'm going to work from home for the rest of the week.'

'OK, no problem,' he said. 'We'll talk on Monday.'

For the next two days, I walked Oscar to school as usual, and then I returned to my PC in the corner of our living room. I buried myself in work, forcing myself to concentrate even while images of a raging Footie Guy flashed through my mind. My insides felt like they were tangled in knots, and my shoulders were stiff and painful, but I'd been pushing through the physical symptoms of fear since I was a young child, and I wasn't going to stop now.

I tried to relax over the weekend, but I was too wound up. Thoughts about what I might say to Footie Guy on Monday kept tumbling through my brain. I felt sick.

On Monday morning, I dropped Oscar at school and then, despite the roiling anxiety, I kept on walking towards the office. I reminded myself how lucky I was, in this big city, to be able to walk to work through Regents Park. I attempted to calm myself by admiring the late autumn colours, and it worked for a bit—just until the point where the white spires of Sussex Place came into view above the treetops.

It was then I noticed a growing pressure in my chest, followed by an increasing reluctance from my legs to propel me any further forward. I consciously willed each foot to move, but with each step the pressure grew until I was forced to a stop, panting for breath and thoroughly confused. Apparently, since I had last walked this way, someone had erected an invisible barrier right across my path, and I couldn't get through it.

I turned back and immediately the pressure lifted. I sat on a bench for a while, wondering what the hell had just happened. After a few minutes I tried again, but the barrier was still there and just as unassailable.

I'd pushed through so much in my life, but whatever this was, it had me defeated.

I again turned my face away from the School and I went home. When I got there, I emailed the manager and said I still didn't feel up to coming in.

I worked from home for another two days before I tried again. The barrier was still there, in exactly the same place, but it had weakened. This time, I was able to grit my teeth and force myself forward. After ten or so wrenching steps, the pressure lifted and only a queasy sense of trepidation remained as I continued onwards.

I went first to Laura's old office to see the temporary manager. I had expected to talk to him alone, but he didn't give me a chance. 'Stay here,' he said as soon as he saw me, 'I'll go and see if he's free now and we'll get this sorted out.' And then he darted out, leaving me off balance and trying to prepare myself for a confrontation with Footie Guy that was coming far sooner than I'd expected.

He returned a few minutes later with Footie Guy, who slouched into a chair and sat there, arms firmly crossed and avoiding eye contact.

As he seemed disinclined to talk, I started doing what I'd already planned, which was to describe my side of the incident to the temporary manager. Almost immediately, Footie Guy cut me off. He charged that I had spoken rudely to him and he accused me of regularly being rude. He said he particularly hated it when I swore.

'But you swear all the time too!' I countered.

'No, it's not the same, you shouldn't do it,' he insisted. I stared at him, shocked into silence by the double standard.

'And that's not sexism, I'm not sexist!' he insisted. Well, it wasn't me who brought it up.

I tried pointing out that I was a good colleague, reliable, always doing what I said I'd do, so couldn't he just overlook anything he didn't like about my personality for the sake of a working relationship?

No, he kept shaking his head, no he couldn't. His body language communicated clearly that *I* was all wrong, that everything was *my* fault.

The temporary manager had sat silently through all of this, not helping at all. I desperately wanted to get through to Footie Guy and to achieve some semblance of a peaceful working relationship, and it didn't look like the manager had anything to offer, so I tried the one trump card I thought I had.

'That phone call...' I started.

While I'd been in Brussels, Footie Guy had occasionally rung me about work things. One time, out of the blue, he had confided some very personal trouble he was going through. I was surprised, but I listened, and I genuinely felt for him. We talked for a long time, during which I shared some of my own past trauma as a way of showing that I understood something about what he'd told me. After that call, I really thought we'd reached a special level of understanding—which made his animosity even harder to deal with after I returned to London.

If I'd hoped that invoking that phone call would bring him back into a moment of connection, I was way off. 'How DARE you bring that up!' shouted Footie Guy. 'You DISGUST me!'

And then he got up and left the room.

Willing myself not to cry, I apologised to the temporary manager.

Yeah, I know, why the fuck was I the one apologising, but I did it anyway.

'Don't worry,' he said, 'I'll talk to him.'

'I want this reported to HR,' I said.

'Yes of course, I'll deal with that too. He can work in another part of the campus today, you just go back to your desk.'

So that's what I did. And like everyone else in the office was studiously doing, I outwardly pretended that nothing was amiss.

A couple of days passed, during which Footie Guy and I kept well out of each other's way. Then the temporary manager came up to me.

'I'm sending you on a communication skills workshop,' he said.

I protested, why me? Why not Footie Guy? He assured me it was just a timing thing. Footie Guy was very busy right now (and I wasn't?) and he would be sent on the next one.

I should have dug my heels in. I had been intimidated and personally insulted, and yet I was the one who needed better communication skills? I mean, I probably did, but surely Footie Guy should have been sent first.

But I was still feeling sorry that the temporary manager had been dropped into this mess, and I didn't want to be a nuisance, so a couple of days later, I went to the workshop. It was a staff development thing, held on campus. There were half a dozen people there, no one I knew.

After we introduced ourselves, the trainer got started. It was what you'd expect: how to listen, how to repeat something back in your own words, the benefits of a follow-up email. I settled in, thinking that maybe this wasn't such a bad way to spend the day, that we can all use a reminder on such things.

Then the trainer started talking about conflict resolution, and my emotions were still too raw. He'd only just begun when I felt a great big sob rising up in my throat. I ran out of the room just as it burst out of me.

The trainer followed me out to where I was fighting to compose myself in the hallway. After a few deep breaths, I told him what had happened and why I was there. He frowned and said, 'It's amazing how often they send the wrong person to these things.' He offered to mark off my attendance and suggested I skip the rest of the workshop and go straight home.

*

In that unsteady way that happens after a crisis, things settled down. Footie Guy avoided me, and everyone else pretended that nothing was wrong. I retreated into my corner, kept my head down and focused on my work—just like I'd learnt how to do all those years before in primary school.

As the end of 2006 approached the CIO announced that this year, for the first time, he was going to hand out awards to particularly deserving IT staff. I wondered if I might get a nod for the significant efficiencies I had introduced in user account management, but there was nothing for me. I wasn't bothered—public recognition would have been nice, but I didn't need it to know I had done a good job.

That all changed when I heard that Footie Guy was getting an award for being the 'Best Team Player'. What a fucking slap in the face.

Furious, I marched straight to the CIO's office.

'How can you give a team player award to someone who has an official complaint against them for aggressive and intimidating behaviour?' I demanded.

The CIO appeared surprised and claimed he hadn't known. According to him, the temporary manager, who had since left, had never mentioned it.

'And he can't have reported it to HR because they would have told me,' he added. 'This makes no sense—he was the one who made the nomination in the first place.'

I was aghast. I couldn't even begin to imagine what bro-code bullshit had been going on between the manager and Footie Guy while I cowered in fear at home.

'I was menaced in the office, and I was so frightened that I couldn't come in for a week—is this really the sort of behaviour you want to reward!!' I spluttered.

'No no, of course not,' the CIO said quickly. 'Leave it with me and I'll look into it.'

So that's what I did. Once again, I left the issue in the hands of a man. And once again, my trust was betrayed.

Footie Guy got his 'Best Team Player' award and, as far as I know, was never counselled, disciplined, or sent on a communication skills workshop. It was clear to me now that Laura had indeed been protecting me—not just from Footie Guy, but from this toxic male tendency to close ranks when 'threatened' by a woman who is just trying to do her best.

I don't want to tarnish all the men in that IT Department, as most of them were lovely to me. However, none of them, to my knowledge, ever took Footie Guy to task over his aggression. This place that had been so good to me for six years now felt hostile.

I took a long break over Christmas and thought about what I should do. We had a mortgage to pay and we needed my income—if I quit, I had no guarantee that I'd find another job for the six months before we left for Geneva. I resolved to tough it out at London Business School a bit longer.

Then, early in 2007, we were slammed with the worst shock of all. Nathaniel's mother rang from Canberra with the terrible news that his father had been diagnosed with cancer, and it was serious.

Now all we wanted to do was drop everything and go to them. My father-in-law had a gruelling three-month course of combined chemo and radiotherapy coming up, and as my mother-in-law didn't drive, it was going to be very difficult. We didn't want them going through this ordeal alone.

Nathaniel's HR department agreed that he could work from the British Embassy in Canberra for a few months.

I immediately handed in my notice to London Business School, and a few weeks later, with no fanfare, my six years with them was over.

It was, I'm almost ashamed to say, harder to leave my MIIS solution than anything else. I had poured so much of my attention, ingenuity, and craft into this thing, and it felt wrong to abandon it. Would anyone else look after it the way I had?

I took one souvenir with me, something to remember my first identity management solution by. Virtual servers were now a thing, and I installed one on my laptop and ported my MIIS solution into it. It wasn't just nostalgia—I had an idea about something I wanted to do over the next few months in Australia, and I was going to need it.

And then Nathaniel and I put our stuff into storage, withdrew our kids from school, rented out our apartment, and flew back to Australia.

21. How to write

IT TAKES MANY HOURS TO TRAVEL FROM LONDON to Canberra. With two long haul flights and a domestic transfer, at least a day-and-a-half of uncomfortable and mostly sleepless travel has passed before you can collapse at your destination.

Fortunately, the kids handled it well. Seven-year-old Oscar was thrilled to have his own screen on the plane, while three-year-old Pascale was small enough to curl up with her head on my lap, and got far more sleep than any of us.

Hours upon hours stuck in a cramped seat gave me plenty of time to think through the idea I'd had in London. I wanted to write about everything I'd learnt about MIIS in the last year, and I had decided to do it in a blog.

At that time, in 2007, blogging was huge. I got a lot of tech news and troubleshooting tips from the technical blogs I followed, and everything from recipes to mood hacks to career development from the lifestyle ones.

My favourite bloggers shared a forensic focus on their topic, but their writing styles were very different. The tech blogs were strictly informative and often dry, while the lifestyle blogs, no less informative, were chatty and personal. I particularly liked Penelope Trunk's 'Brazen Careerist' and Gretchen Rubin's 'The Happiness Project' for the way they obsessively examined their topic from all possible angles, framing their thoughts within stories from their own lives.

I wanted to write about MIIS and identity management in a way that combined both styles. Not just 'How to do X', but 'Here's why I had to care about X, and this is how I figured out what to do about it'. Useful technical content wrapped in a little story.

I had a name for my blog too: it would be called *Miss MIIS*.

*

I wouldn't say writing comes easily to me. At school I struggled to corral my thoughts into the strictures of essay writing, and my fiction was flabby and derivative—but I never shied away from writing, and this would turn out to be incredibly helpful in my career. In all thirty-five years of it, I doubt a week has gone by when I haven't had to write some sort of document for other people to read.

At high school, florid and verbose compositions that demonstrated your capacious vocabulary were rewarded with exemplary ratings—but that's no good for tech writing which needs to be clear, succinct, and unambiguous.

It was my time working at AusCM in Sydney that really honed my tech writing skills, but I'd had a prior introduction to it at university. It was in a Computer Science elective called 'Computing Milieu', and it was, hands down, the most useful one for my career.

This class was not popular. I only took it because my timetable was complicated by my tutoring work, and likewise, scheduling issues seemed to be the only reason anyone else was there. Uptake was probably not helped by the fact that none of us knew what 'Milieu' meant.

This course was a passion project of the lecturer, and he'd been trying to get it on the electives list for several years. The year I took it was the first it ran and, very sadly I think, also the last. I guess a lot of people didn't get what he was trying to do: while our other classes were all about *doing*, this lecturer wanted to teach us to *read* and *write* and *think* about computing. And you can't, I maintain, be a good IT pro without all of that.

In Computing Milieu we read *Sleepers, Wake! Technology and the Future of Work,* by then Australian Minister for Science and acclaimed polymath Barry Jones. I found the book fascinating. Until then, I'd thought about computers only in terms of what they could do, but this book made me ponder the wide-ranging societal consequences of technology. I was eager to discuss it in tutorials, but I seemed to be the only one. My classmates grumbled about having to read the book, and they were even more unimpressed when they heard we had to write an essay about it. 'I didn't take Computer Science to write *essays*!' they complained.

Our other set text would, our lecturer assured us, help with the essay writing. It was the classic on concise writing, *The Elements of Style* by Strunk and White, which argues the benefits of writing with simple language and clear phrasing—a 'less is more' ethos perfectly encapsulated by the slimness of the volume.

Of all the textbooks I read in my degree, this is the one that's helped me the most. When writing, I still remind myself to 'omit needless words', and 'keep related words together', and 'use specific language'.

I've worked with many colleagues who are extremely reluctant to write, and I get it. Writing clearly and with just the right level of detail is *hard*—but it is an essential skill in tech. AusCM, for example, wouldn't have had a business without good writing. We had to write bid responses and project plans and designs and reports and change requests and instructional guidelines, and they had to be both stringently correct and legible enough to be quickly scanned by busy people. At AusCM, while my title was Technical Consultant, half of my job was writing.

Is AI changing this? I don't doubt it will help with boilerplate content, with phrasing and with smoothing off the rough edges, but it's not going to do it all. I think it will remain just as important for tech workers to communicate clearly with the written word, whatever tools they use to help them.

*

We arrived in Canberra at the end of April 2007. My in-laws, Tony and Doreen, didn't have enough room for the four of us, but my old school friend Emma, the one whose house I used to flee to every time my stepmother kicked me out, now lived in Canberra. Once again, she was kind enough to offer me a port in the storm, and the four of us moved in with her, her husband, and her two boys.

Oscar would be going to school with Emma's sons, and Doreen had already arranged playdates for Pascale. Nathaniel would be going to work at the British Embassy, and I would drive Tony to and from his appointments. I'd be doing a lot of waiting, but I planned to make full use of it.

THURSDAY 10TH MAY 2007, CANBERRA

This drive has quickly become familiar, though that doesn't stop Tony, my father-in-law, directing me as I take him to the Canberra Hospital. He's a couple of weeks into a combined course of chemotherapy and radiotherapy, three gruelling sessions a week.

He's keeping his spirits up, and as we drive, he discusses Richard Dawkin's book *The God Delusion*, which we've both recently read. It comforts me that this stark confrontation with his own mortality has not changed Tony's determinedly rationalist opinions one bit.

What is not comforting is how painfully thin he has become since the cancer affected his ability to swallow. Doreen, my mother-in-law, has kept him supplied with protein shakes, and I've been making him soft, slow-cooked stews accompanied by buttery couscous that he can still get down. Despite our efforts, this dear man is wasting away in front of our eyes.

I drop him off outside Oncology before I hunt for a parking spot. As he gets out he says, as always, 'I'm so sorry to be wasting your time like this.' And I reply, as always, 'Don't worry about me, I'm working on my project'.

Soon I am sitting in the oncology waiting room, along with the other people waiting for their loved ones. Most are reading books or flicking through magazines. I am the only one balancing a laptop on my knees.

I have no internet connection, but I don't need it. I'm drafting the blog articles at the moment, and I'm not going to post them until they're good and ready. I've mapped out fifteen posts I want to write, and today I'm working on number seven: 'A Cheater's Guide to SQL DTS Packages'.

My goal for today is to add screenshots. I fire up my little virtual server that hosts my MIIS solution, and while it's booting, I review my post.

I tweak a word here and there while thinking, as always when editing, about Strunk and White. They might not like my use of jargon, but I think they'd approve otherwise:

> Let me just say that SQL was a mystery to me before I started working with MIIS. I'd always looked at it from a sys admin point of view–install the server, maybe do some patching, and then hand it straight over to the database people.

> But you can't do MIIS without learning some SQL and I soon was bugging my friendly, neighbourhood database administrator for advice on query scripts and how to go about running them. According to her, I needed something called a DTS package.

My virtual server is running now and I click through the DTS packages in my solution, taking screenshots and cropping the images. I'm so absorbed in what I'm doing that I don't see Tony until he's standing right next to me, ready to leave.

'The nurses are so great,' he says, as I jam my laptop back into my bag, 'always joking with the patients. I never thought having cancer could be so funny!'

I want to hug him as hard as I can and never let him go, but that's not really Tony's style, so I just nod and smile and hope for the best.

*

In late May, I toggled my first post to 'public' and added a blog link to my signature on the MIIS TechNet forum. Nothing happened straight away, but as I continued to publish a new post every few days, I got hits. Not huge numbers, but there were definitely people looking at *Miss MIIS*.

I quickly learnt what all content creators do: checking your statistics is way too addictive. That hope when you open the stats page, that flush of validation when the numbers go up, that disappointment when they don't... and then checking again five minutes later in case it's changed. I had to make a stern deal with myself: no looking at stats unless I had just posted something new. A dopamine reward for my self-assigned work.

Within a month, I had published all the posts I'd written in hospital waiting rooms and I had maybe fifty readers. A modest beginning, but encouraging nonetheless.

A few people had been nice enough to send encouraging messages. Back then, we only had official MIIS documentation, and it ranged from frustratingly sketchy to impenetrably detailed. My blog was the first to offer real world examples in plain language. Once again, as I'd done with MIIS on the Exchange project, and going back to my days as a grunge violinist, I had found a niche of my own.

I would go on to write regular *Miss MIIS* posts for years to come, at least two a month with original content and a lot of thought. This is not going to turn into a story about me becoming a blogging superstar—MIIS was never mainstream enough for that—but it did make me well known within my specialist community. It also led to me receiving a great honour that kicked off a new phase of my career, which we'll get to later.

In Canberra, Tony's treatment cycle ended, and at the beginning of August it was time for Nathaniel, the kids and me to return to Europe and Nathaniel's next posting in Geneva. Saying goodbye was hard. We hoped the treatment had been successful, but as I hugged Tony, he felt so frail that something in me knew it was for the last time.

Tony Wapshere, entomologist, bon vivant, uproarious laugher at his own terrible jokes, and all-round decent human being would leave us only a few months later.

When that day came, we were settled in Geneva. Only Nathaniel could return to Canberra for the funeral, but the kids and I honoured Tony's memory as best we knew how. And we have continued to do so ever since in a way that is universally human, while at the same time unique and precious.

We tell stories.

22. How to arrive

BEFORE NATHANIEL'S JOB MOVED US TO GENEVA in August 2007, I'd only been there for the 1998 Internet conference, a couple of weeks before my fateful encounter with him at his 30th birthday party. All I remembered from that visit was the convention centre near the airport, and that it had been so swelteringly hot that some wag quipped it was the only time he'd lost weight at a conference. I had seen very little of Geneva and was unaware that the haze which settles in warmer weather had completely obscured the view of the Alps. I hadn't suspected the mountains were so close.

So many things about that 1998 trip looked, in retrospect, like signals of fate. The email from Diana arriving at just the right time; my AusCM colleague joking that I shouldn't go to the other side of the world and fall in love with someone I already knew; my friend in Toulouse mentioning that Nathaniel was newly single; how of all the possible conferences, I chose one in Geneva. Logically, I knew my human brain was ascribing meaning to chance events, but I couldn't deny a sense of destiny as we drove from the west, through France, and towards the Swiss border.

We got our first glimpse of Switzerland as we came through the pass at the top of the Jura range. A view of the Alps, the largest mountain range in Europe, unfolded before us. We stopped at a lookout to gaze open-mouthed at the snow-covered peaks stretching into the distance. Towering above them all, magnetically drawing the eye, the mighty Mont Blanc loomed like a deity gracing us puny mortals with a glimpse of its majesty. I had grown up in a country where the highest mountain would barely be considered a hill in the Alps, and the sight of those endless, white behemoths was too much to take in.

Eventually, I wrenched my eyes away and looked down at the land between the Jura and the Alps. To my left the broad expanse of Lac Leman swept into the distance. To my right, barely discernible through the low-lying summer haze, I could just make out the city of Geneva, clustered around the southern tip of the lake. Our new home.

*

Our first few weeks in Geneva passed in a flurry of activity. Nathaniel had to start work immediately, so it once again fell to me to set up the house, register us for banking and medical services, settle the kids into their new home, and get them ready to start school. Looking for a job was also on my to-do list, but it was way down at the end.

September rolled around and the kids went off to school. I could now turn my attention to job hunting.

A specialist job with MIIS was my ultimate goal, but in 2007 the concept of automated identity management was still in its early days, and I knew that was a long shot. Instead, I hoped to find a general tech consulting job that I could perhaps bend towards identity management, while at the same time keeping up my other skills.

As usual I trawled internet job sites, but this time, I also had another thread to pull. The way I'd got the job at London Business School had shown me the power of a personal introduction, so a few months earlier I'd signed up to a new website that was all about who you knew. Now I planned to mine my LinkedIn network for Geneva connections.

My network was small—mostly ex-colleagues from London Business School and a few vendors—but even so, I found a possible lead. A couple of my ex-colleagues had connections at the Lausanne School of Business, only 40 km up the lake from Geneva. I sent some messages, and before long, I had an invitation to meet their IT Services Manager.

On a sunny morning in mid-September, I drove to Lausanne for my meeting. The IT Manager was very welcoming, showing me around the School with noticeable pride, before taking me to the cafeteria for a coffee. He revealed that there had been quite a bit of collaboration between the IT Departments of the London and Lausanne Schools of Business, and that he'd spent some time in London, shadowing our IT managers and learning how we did things.

'I can't offer you a job myself,' he said, 'but I'm very pleased to help someone from London Business School after all the help they've given me.' It was touching to benefit from this good will that my former colleagues had earnt.

He offered to forward my CV to a few consulting firms he dealt with. He must have done it straight after I left, as my phone rang that very afternoon with a call from an IT Services company called LANExpert. This call set off a rapid sequence of events over the next few days.

First, I was invited for an informal meeting with one of the Directors, a twinkly eyed older gent with whom I had a lovely chat. This led to a formal interview the very next day with a couple of the other Directors.

As part of this interview, I was asked to show on a whiteboard how I'd migrate users from an old directory system to a new one. I drew servers and scrawled arrows and talked through my thought process, referencing the Exchange migration I'd done at London Business School. They could see I knew what I was talking about, though they were a little surprised when I included a MIIS server in my diagram.

'We haven't seen much demand for MIIS here,' one of them said, 'but it's very good that you have recent Exchange experience. That is very interesting to us.'

I left with high hopes. The interview had all the hallmarks of a good one: the interest in my thought process, the mention of what gave me an edge, the 'see you soon' when I left. And I was right, a job offer came through that very afternoon. It was only three days since my coffee with the IT Manager in Lausanne—it was clear these people didn't muck about!

Back in the LANExpert office the following day to sign my contract, I noticed something interesting. My interviews had been conducted in English, thank goodness, but I had enough French for social niceties. I had been addressed with the formal 'vous' by everyone I'd encountered, but as soon as I signed the contract, they immediately switched to the informal 'tu'. It was a linguistic 'welcome to the family', and I took it as a good sign.

There was another good sign after I returned home. Again, the phone rang with a different consultancy firm offering me an interview.

'I'm sorry,' I said, 'I just accepted an offer from LANExpert.'

'LANExpert?' sighed the caller. 'You would have been perfect for us!'

*

The good signs proved right. At LANExpert, I joined a cadre of technical guns for hire who were among the best and brightest people I have ever worked with. LANExpert also proved to be the best company I'd ever worked for.

The company had been founded by former IT consultants, and that goes a long way to explaining what made them so great. They understood what they were sending us out to do, and they set us up for success.

I could go on for days about their sensible hiring practices, their fair and reasonable employment and remuneration policies, and the way they actively built a company culture among a workforce that was scattered across hundreds of customer sites. I sensed a clear intentionality in the way they had designed the company, and it was based on values they gave far more weight than mere lip service.

Only a few weeks in, after my manager called me into his office, I understood just how fortunate I was to have landed with LANExpert.

'We have a customer who wants to build a new Exchange 2007 email system,' he said, 'I know you have worked with Exchange 2005, but not 2007—so how do you feel about taking this assignment?'

I had read enough about the recently released Exchange 2007 to know it was architecturally very different, and I'd have a lot to learn. But that wasn't anything new. I told my manager I was up for it.

He asked me to stay while he phoned the customer. Now, if this call had gone how I—based on past experiences—expected, I would have heard him say something like this:

'I understand you're looking for someone with Exchange 2007 experience? Well, you're in luck! One of our highly experienced consultants has just become available, and they can start straight away!'

And I would have sat there silently, with a queasy anxiety snaking around my belly, knowing that a false expectation had been set, and that I had been trapped into responsibility for it.

Instead, he said:

'I understand you're looking for someone with Exchange 2007 experience? We have a consultant, however he is fully engaged and his next availability is in two months—I can book you in then if you like. Alternatively, I have someone with excellent Exchange 2005 available immediately. Which do you prefer?'

That I was stunned should tell you something about the pervasiveness of embellishment, fabrication, and outright lying in commercial IT. There had been no hard sell and no misrepresentation. Just the simple facts and a simple choice.

The customer elected to have me come straight away, and I can't tell you how freeing it was to be able to admit what I didn't know, and to openly look up documentation, secure in the knowledge that no one had been misled.

Since then, whenever anyone tries to tell me that a few 'white lies' are just the unavoidable realities of 'doing business', I call bullshit. LANExpert's business didn't suffer from their honesty, in fact it thrived.

*

Geneva has to be one of the most interesting places to be an IT consultant, punching well above its size with the variety of organisational sectors. I assisted customers from banking, health, education, manufacturing, trade, and charity. There were multi-nationals, local businesses, public services, and the United Nations agencies. One day I'd be at the Union of European Football Associations, the next at some fancy watch makers, and the following at the Geneva bus depot.

I worked in many a nondescript, cubicled office, but also in some memorably different ones. There was the IT Department banished to a creepy old house in a forest, and the exclusive private school in a Hogwarts-esque castle on the edge of the lake. I repaired an email server in a mechanic's garage while sitting next to a girly calendar, and a domain controller in the offices of a major skin-care brand while sitting next to a life-size cutout of a famous model in her underwear. At the bus depot, the IT Department was above the maintenance shop and on my way up I'd see buses undergoing major surgery, engine parts arrayed on the floor around them. At a wealthy firm in a prime location, I kept being distracted by the spectacular views of the mountains and the lake, and I wondered how anyone got anything done.

I was fascinated, not just by the surroundings, but by how differently my customers used their IT. As a

Microsoft-centric consultant, I saw the same tech everywhere, but the way it was used could be surprisingly unique. One company had built their core business processes around Exchange public folders—a little-used feature elsewhere—and had 10,000 of them with more being added every day. An international charity needed built-in failsafes for *when* (not if) a server got stolen from an aid camp.

I loved the variety, but the itinerant work schedule took some getting used to. At first, I approached every new engagement with trepidation. What was I walking into? Would I completely fail to fix the problem, or even make things worse? Would the customer take one look at me and demand someone else be sent, someone better, anyone but me?

I pushed through, and over time, as I successfully completed one engagement after another, and as customers were friendly and complimentary, my nerves subsided. A few months in, I reached the point where I looked forward to a new customer. I now enjoyed meeting new people and hearing about their tech issues, and I was curious to see how I'd do.

There are many stories I could tell you, but I've chosen this next one because of the confidence boost it gave me. As someone who always set myself a high professional standard, this was a moment when I knew I'd arrived.

TUESDAY 24TH JUNE 2008, GENEVA

I'm perched on the edge of a padded bench, waiting for someone to come and collect me from reception. Above me is a lofty open space, a four-storey-high void, bright with sunlight that streams through tall windows. High on the wall, an elaborate clock is mounted. Pointed rays stretch out from the circumference of the clock face, and it glows golden in the light, like a messenger sent by the sun itself to tell us the time.

I'm in the headquarters of a famed manufacturer of elegant and very expensive timepieces. I've never visited this company before, but when their urgent call about a malfunctioning database server came to LANExpert this morning, I was available for immediate dispatch. So here I am in this impressive atrium, waiting for my customer.

My gaze drifts again to the vault of space and light above.

Wait a sec… I check my own cheap watch, and then, to be sure, my phone. There is no doubt. In this cathedral to the precision of Swiss watch making, the great sun clock is twelve minutes slow.

The customer arrives and, as we walk to the server room, he explains the problem. They have two database servers that have been set up in a cluster. They are supposed to mirror each other, with one taking over if the other one fails, but something has gone terribly wrong.

'We had to restart one of them yesterday and since then, they keep bouncing,' he says. 'One comes up and the other goes down, then it comes up and the first one goes down, over and over again. It's been working fine for months until this. Many other systems use the database cluster and they're all broken too.'

I don't know much about database clustering, but his description is ringing bells. It reminds me of a time at London Business School when something similar happened to our network storage.

The customer leads me to a terminal and asks me to wait while he logs in. Standing behind him, I'm still thinking about that night in London. I'd been on call, so when our servers started crashing all over the place, I was the one hauled out of bed to deal with it.

I had logged in to the School network from home and quickly identified our network storage system as the culprit. Like this customer's database server, our storage system was made up of two mirrored nodes, with one able to step in should the other fail. But on that night, they had started fighting for primacy. It was just like this customer described, with the nodes 'bouncing' and the service unable to stabilise.

I had phoned Dell, our hardware vendor, for assistance, and then spent most of the night on the call with a series of support technicians. I had watched as they scoured log files and tried different settings–it was a five-hour crash course in cluster configuration. At around three in the morning, after the latest support tech's shift had ended and I was waiting to be passed on to the next one, an idea had surfaced in my overclocked brain.

When Dell Tech #4 joined the call, I was ready for him. 'I want to increase this timeout here. I think the nodes don't have enough time to see that the other one is the primary, so they keep trying to grab the role. What do you think?'

'Uh yeah, sure,' said the tech.

I upped the timeout setting, rebooted the nodes–and the cluster stabilised.

'Oh, yes, that's obvious really,' said Dell Tech #4, no doubt wondering why the call had been open for five hours and transferred three times when the fix was so easy. But that's troubleshooting for you: the really hard work is finding the problem.

Back in the present moment, the customer has logged in to the cluster admin interface and is gesturing for me to take his seat. Normally I'd start with a deep dive into the error logs, but I figure, what the heck, maybe I'll just have a quick look and see if there's anything like that Dell timeout.

Scrolling through the long list of settings, I spot a timeout which looks familiar. I increase its value and then, not even daring to hope it could be this easy, I restart the servers.

The first one to boot becomes the primary node, and the second one quietly settles into its secondary status. No seizing of the primary role, no bouncing, no chaos.

I confirm the fix by again rebooting each server, and the cluster behaves perfectly. In no more than fifteen minutes, I have solved the problem.

The customer is delighted. He spends a few minutes confirming that the other impacted services are coming back online, and then he walks me to reception. He is relaxed now with the crisis over, and he chats in a friendly way, pointing out features of the building and telling me some of its history. In the atrium, I feel emboldened to ask him why the sun clock has the wrong time.

'Oh yes,' he says, frowning a little, 'it's been running slow for years. I don't know why they don't just fix it.'

'I guess the timing was never right,' I joke. He chuckles politely, like he hasn't heard that one before.

I take my leave and walk out into the beautiful spring day. I feel like jumping into the air and pumping my fist in victory, but there are people about, so I limit myself to a happy grin.

I'll be turning 40 in just over a year. I used to dread that age, but now I'm feeling fine about it. I know I got lucky today, but also, that it wasn't just luck. I solved that problem because I have something you can only get with time: I have experience.

*

As I went from one LANExpert customer to another, I often reflected that my decision to stay technical had paid off in spades. If I'd given in to those mid-thirties anxieties and switched to Project Management I would, without fluent French, have been almost unemployable in Geneva. Instead, as an experienced techie, I was in demand. I may have had only patchy French when I arrived, but LANExpert gave me a chance because I was fluent in the international language of IT.

I later heard that among the 'trailing spouses' of the British diplomatic community, I had a reputation for the fastest zero-to-employed of any spouse that anyone had heard of. If I'd needed more proof that staying on the technical track had been the best professional decision of my life, this clinched it.

At LANExpert, I worked with technical consultants aged in their twenties to sixties. Not since those crusty old dinosaurs Civil and Mechanical, the ones who had so blighted the final year of my first IT job, had I worked with techies in the over-fifty age bracket—but where Civil and Mechanical had been a warning, these men were a beacon. They were still engaged, still adaptable, and still great at what they did.

As I got to know these older men I asked if they'd ever worried about aging out of tech work, and they seemed surprised by the question. They all said much the same thing: they were good at what they did, they liked it, it paid well, why stop?

Why indeed?

23. How to be exceptional

ONE DAY, WHILE ON-SITE WITH A LANEXPERT customer, I was struck by a thought. If you're a demographic minority, you have to find your own path while others simply follow the crowd.

The customer was a university, one that specialised in technical degrees such as Engineering and Computer Science. They are located deep in the French-speaking part of Switzerland, a good hour's drive north of Geneva. I had been engaged to help them build an identity management solution, and this was my first day.

After a morning spent discussing requirements, the team took me to the university's cafeteria for lunch. We all ordered the *plat du jour*, and then I followed them to a table reserved for staff in a raised corner of the dining hall. I looked out across the sea of chatting and laughing students, packed in together on long benches, and from a habit borne out of countless times in male-dominated spaces, I played 'spot the girl'. There were at most one in fifty.

'Where are all the girls?' I asked my lunch companions.

'They're down the road at the School of Business,' one of them said, 'hardly any want to come here.'

Another added, 'Our school spends so much money on programs to encourage them to study Engineering or Computing, but it never changes.'

We all shrugged, shaking our heads at the intractability of the situation, and the conversation moved on. It was only after lunch, while walking back to their office, that something occurred to me.

How many of those students had no particular passion for their degree, but had simply followed a well-trodden path to this school? Some kids finish high school with a burning conviction about their future, but most of us, myself included, do not and we follow family expectations or the examples of our friends or what our teachers tell us. I may have ended up in an unusual career for a woman of my generation, but in selecting a Bachelor of Science degree, I was following exactly what my father and teachers had told me to do. Even Computer Science was, back then, not unusual for a girl to choose.

Now we were two decades into the downturn of women in computing, and the students in that cafeteria had grown up with a very different set of social expectations. I imagined that for the boys, a wide and well-trodden highway led to this university, while for girls, it was a stony and barely discernible goat track.

How much passion and conviction must those girls have had to find their way here? How exceptional must they have been?

*

A year later, a recent graduate of that school started at LANExpert. Until then, I had been the only female consultant among a hundred men. Now, as Marie Thérèse joined us, I became one of two.

Marie Thérèse was born and raised in a former French colony in Africa. She was very bright, good at maths and science, and had been encouraged to apply for a scholarship to a European university. Switzerland accepted her, but as the quota for her first choice of a medical degree was already full, they invited her to study *Informatique* instead.

She did not know that *Informatique* meant Computing. She'd never even seen a computer and thought it might have something to do with books, and she liked books, so she accepted. It was only after moving to Switzerland that she found out what it was all about. She had started far behind her classmates who had grown up with computers, yet four years later, she graduated top of her class.

'What was it like in that place,' I asked her, 'being so far from home and looking so different to everyone else?'

Marie Thérèse just shrugged, and smiled her beautiful warm smile, and said it was fine.

I marvelled at her resilience and her tenacity. There must have been hard times and many obstacles in her path, but she had persevered. I had wondered about the few young women in that cafeteria and now I had met one. She was undoubtedly exceptional.

*

I was accustomed to being the exception in IT purely for my gender, but I didn't like to dwell on it. Except in

Switzerland, where I had little choice. Working there, I was more regularly reminded of my gender than I ever had been in Australia or the UK.

First, there was the kissing. Not long after I started at LANExpert they held a celebratory company dinner, and as I hadn't yet met many of my new colleagues, one took it upon himself to introduce me around. Because I was a woman, every 'Enchanté' was followed by three cheek kisses. A hundred and fifty kisses later, I thought, 'Boy, it's exhausting to be female around here!'

And then there was my name—my francophone colleagues and customers could not deal with how I spelled it. I signed off emails as 'Carol', and they replied with 'Chère Carole'. In French, you are constantly reminded of your femaleness with a whole lot of supplementary Es.

The other thing I noticed was the curiosity. People came right out and said, 'It's very unusual for you to be doing this job, I've never met a woman doing technical work before.' Sometimes they asked if there were a lot of female techies in Australia. Perhaps they imagined I came from a land down under where women rode a kangaroo into the office, tackled a few server issues with a giant screwdriver, and then wrestled a croc on the way home.

I liked it when colleagues or customers admitted their surprise at my non-standard gender for this job. It never felt disrespectful or judgemental, and it gave me a good opportunity to chat with them about it. And as I learnt more about Swiss culture, I understood where these comments were coming from.

Switzerland, I had been shocked to learn, had one of the worst track records on gender equality in the OECD.

For instance, Swiss women only gained the right to vote in federal elections in 1971 (compared to 1901 in Australia), and one canton prevented women from voting in local elections all the way up until 1991!

And of course getting the vote is only part of the struggle—Swiss society was still very traditional when it came to gender roles, particularly with parenting.

I became aware of this when LANExpert's much-loved receptionist returned to work after maternity leave—and then, only two months later, resigned.

'It's impossible,' she told me when I asked why she was leaving, 'there's no childcare that I can afford, and even though my mother helps as much as she can, it's not enough. I'm so sad because I love working here, but I have no choice.'

I thought about what I would have done if I hadn't been able to find good quality childcare for my kids in London—would I even have a job now, let alone one in my chosen career? It didn't bear thinking about. It seemed completely bizarre to me that a woman was being forced into a choice like this in 2008!

And she wouldn't have found it any easier once her children started school. I was lucky to send my kids to the International School with its accommodations for working parents. If they'd gone to the local school, I would have had to drop them off at 9 am, pick them up at 11 am for lunch, return them at 1 pm, and then pick them up again at 3 pm. Holding down a job would have been impossible. In Switzerland, it wasn't just working in IT that made me an exception, it was also that I was doing it as the mother of primary school aged children.

And yet I never noticed any judgement or bias—as far as I could tell, my colleagues respected me and my customers appreciated my work. It is only in looking back now that I see the lengths I was going to, not just in Switzerland but throughout my career, to win that respect and appreciation. I had long since accepted that the automatic trust my male colleagues were accorded for simply showing up would never be mine. Proving myself, time and time again, was just part of the price I had to pay to have this career.

On a subconscious level I understood that because I was an exception, I had to be exceptional.

And this is a problem, not just for me, but for the whole tech industry. While female enrolments in computing are finally increasing, they have yet to reach the thirty percent level of the mid-1980s. For young women, it is still a narrow path which they must be exceptionally determined to follow.

For a technically-minded woman in an IT workplace, the risks of being anything less than exceptional are still too high. Women are judged more harshly than their male counterparts while their achievements and contributions are minimised and overlooked, and the direct result is a quit rate for women that continues to be the highest in any STEM profession. Fifty percent of women who start a career in tech leave by their mid-thirties.

It's fine to be exceptional, but it shouldn't be a prerequisite for only a subset of the workforce. As I've said many times, only half-jokingly, we won't reach equality in IT until there are just as many mediocre women as there are mediocre men. A more polite way of putting this is that the path into and through this profession should not be systemically harder for one demographic group than another.

My own path through this industry had been rough at times, and I'd come close to losing my way, but I was still here. And with my fortieth birthday just around the corner, I was about to be rewarded for my tenacity in a way that I hadn't even dreamed could be possible.

24. How to reach your peak

I LOVED WORKING AS A GENERALIST CONSULTANT for LANExpert, but I didn't give up on my goal to become a MIIS specialist. Every now and then, identity projects came along, and knowing that was an interest of mine, my manager assigned them to me. And whenever I did anything involving identity management or MIIS, I blogged about it.

I wrote about specific requirements I had been given, along with detailed step-by-step instructions on how I met them. I wrote about faults I had encountered and what I did to troubleshoot them. And I wrote a lot of my thoughts on what constituted good and bad design for an identity management solution.

In mid-2008, about a year after I'd published that first batch of posts written in hospital waiting rooms, I received an email from someone in the MIIS product group at Microsoft.

'Hi Carol,' read the email, 'I've noticed you answering questions on the forum and your responses are always very informative. I've just had a look through your blog, and I'd like to nominate you for an MVP, if that's OK with you.'

He was offering to put my name forward for a Microsoft Most Valuable Professional award. MVPs are awarded annually to around 4000 people worldwide who freely share their outstanding expertise with the IT community. You will find MVPs speaking at conferences, answering questions on forums, running open-source code projects, organising user groups, hosting a tech podcast, and doing any number of other unpaid activities which educate and assist their fellow IT professionals.

I'd sat in conference sessions presented by MVPs, and I followed quite a few MVP blogs. To me, MVPs were the true rock stars of the tech community. You didn't get a hollow sales pitch from MVPs, you got deep technical knowledge tempered with a lot of real-world experience, and I'd rather listen to that than a marketing spiel any day.

I'd even met some of the small number of MIIS MVPs. Another of LANExpert's fantastic employee benefits was a personal education budget, and I'd chosen to spend mine on going to The Experts Conference, a highly specialised three-day event in the USA. It was the only conference with an entire track for MIIS, and it attracted the hard-core obsessives. Surrounded by my people, with no conversation too obscure or in-depth, I was in heaven. And the coolest kids in that incredibly nerdy school were the MIIS MVPs. They dazzled me, and I felt honoured that they so readily welcomed me in.

At my next review with my LANExpert manager, I admitted a secret dream I'd had since meeting those MVPs at The Experts Conference. My manager had just asked if there were any certifications I wanted to do, and I said, 'I've done a lot of certifications in my life—what I really want now is an MVP.'

'It's difficult,' he said, 'it's not like you can do an exam or apply. I don't even know how they decide who gets one. Do you?'

I had no idea either—not until, some months later, the email arrived from the product group guy. I gleefully accepted his offer to nominate me, but then I heard nothing for a long time, and I wondered if he'd forgotten. Until, on the 1st of January 2009, I woke up to find an email waiting for me:

'Congratulations 2009 Microsoft MVP!'

I was so excited that I ran around the house, jumping up and down while my kids laughed in surprise, and Nathaniel emerged from our bedroom wanting to know what all the fuss was about so early on New Year's Day.

'I'm an MVP!!!' I shouted, dancing on the spot.

After many hugs I collapsed, grinning ear to ear, into an armchair. My daughter, too young to know anything other than something wonderful had happened to Mummy, snuggled happily on my lap.

That MVP award was the greatest affirmation of my professional life. It felt like all of IT had stood up and said, 'Carol, you belong, you are appreciated, you count'. And it was all the more meaningful because I got to share that moment with my loving family. That bullied, neglected, and abused kid I had once been, that little Smellbomb who still lived inside me, had made it, more than she'd ever dare dream.

Two weeks later I received a perspex MVP trophy (which went on the mantlepiece), an MVP lapel pin (which I never wore), and permission to put 'Microsoft MVP' on my blog and business card (which I absolutely did). I wonder how many other MVP awardees still use the blue padded box the trophy arrived in to store their sewing stuff!

Now I was an MVP, speaking invitations followed. My favourite nerd heaven, The Experts Conference in the USA, offered me a spot, and I went on to present sessions there every year until it folded in 2012. I spoke at Microsoft TechDays in Geneva, and later, after we moved back to Australia, at Microsoft TechEd in Australia and New Zealand. Preparing and rehearsing my sessions was a huge amount of work, and I was a mess of nerves before speaking, but once I was up there on the podium, I enjoyed myself. I guess I still had enough of that former grunge violinist in me to relish a little bit of the spotlight.

*

In July 2009, seven months into my tenure as an MVP, I turned forty.

That age I'd once dreaded was instead the pinnacle of my career. I loved my job with LANExpert, I was a well-regarded consultant and now a Microsoft MVP as well. All my struggles, my pivots, my courage, and my determination had brought me here. I felt like I'd spent twenty years climbing Mont Blanc and now here I was at the top, breathing in the clear air and revelling in the glorious views.

However summiting the peak is not the end—from the top of Mont Blanc you still have a long journey and many obstacles ahead of you. At least now my path forward looked clear. I would continue to do any identity projects that LANExpert could give me, and when the time came to leave, I would find that specialist role.

Before that, we had two more years in Geneva. We made the most of it. We played in the snow in the winter and camped in the high alpine meadows in the summer. We took full advantage of our proximity to France, doing all our fresh food shopping at the weekend markets, searching out the most intensely perfumed cantaloupe or the stinkiest cheese. Some weekends we drove through the long tunnel underneath Mont Blanc, straight into northern Italy with its alpine villages and heady truffle markets. And on evenings when Mont Blanc emerged from its usual haze of clouds, we'd walk to a nearby vineyard with an unobstructed view of the lake and the mountains. Those were among my favourite times, standing together with my family and basking in the awe of the geological spectacle before us.

We loved living in Geneva, but we never forgot that we were in a privileged bubble, and that it would deflate one day. Nathaniel's posting was due to finish in mid-2011, and we had decisions to make.

Our two options were another posting or a return to London—until a third presented itself.

Nathaniel, in his work with the British delegation to the UN, had befriended some of the Australian diplomats, and it had been suggested to him that he might apply for a job in Canberra. I was surprised when he said he was seriously considering it, and by the jolt of hope this gave me.

My first couple of years in London had been rough, what with the homesickness and the dot-com disappointments. Since then though, I had found much to love about living in Europe. I had accepted a future in which, wherever Nathaniel was posted, London would officially be 'home'.

But I had regrets about what I had missed back in my real home, and top of that list was my half-siblings' teenage years. My sister, now all grown up, had visited us in Geneva for my fortieth birthday. She'd become a warm, funny, and accomplished young woman and I adored her, as did Oscar and Pascale. It was bittersweet to watch how much they enjoyed having their cool young auntie around for those few days. During my childhood, all my extended family had lived in New Zealand, and I only saw them on occasional trips across the Tasman. It saddened me to see how I had repeated that pattern with my own children.

When Nathaniel suggested we apply for jobs in Canberra, it released thoughts I had been keeping locked away. I acknowledged then how much I wanted Oscar and Pascale to know their family and, more broadly, to know Australia. They had spent their lives in Europe, and Australia was just this place we visited occasionally on holiday. I wanted them to have that connection to country, that anchoring you only get from growing up in a place. I wanted them to call Australia home.

Nathaniel too had a deeply personal reason to go back. His mother Doreen had been alone since Tony died, so when we made our decision to move to Canberra, it was as much for her as anyone else.

Being, as I was, at the peak of my career, the next job search was the easiest of my life. I simply reached out to an Australian firm that specialised in MIIS solutions and let them know when I'd be moving to Canberra, and an offer to become one of their principal consultants soon followed. I won't deny it, after twelve long years away, I felt like a conquering hero returning home.

25. How to work in a meritocracy

I STARTED WORKING FOR SYNCID RIGHT AFTER WE moved to Canberra in mid-2011. They were a relatively new Australian company, and their core business was custom identity management solutions built on MIIS. They seemed to be doing well with offices in the major capitals, including Canberra, and an impressive list of customers.

They were clearly proud to have me join the company. At that time, I was the only MIIS MVP in Australia, I was a *name*. In my first few weeks I was taken around to various customers and introduced as 'Carol Wapshere, *our MVP*' like I was some kind of talisman.

Only the CFO seemed less enthusiastic about my arrival—not because he didn't like me, but because they didn't have billable work for me right away. I was earmarked for an upcoming project, but it was still two months off, and in the meantime, he was paying me to shake hands. And I agreed with him, because in commercial IT you're only as good as your last customer invoice. Whatever reputation I'd come in with, I still needed to prove my merit.

*

For the longest time, I believed that I worked in a meritocracy. 'In IT', I'd say, 'all that matters is whether you can do the work, not who you are or what you look like.'

Sure, there had been times when I'd been treated with such blatant disparity, I couldn't ignore it—like that whole Footie Guy situation where he was rewarded while I was discounted and blamed. Another was a project I did in the 1990s where the PM criticised me on such a regular basis, and with such scorn, that my male colleagues actually noticed. They thought it was funny and turned it into a game, deliberately repeating whatever I'd just said and been rebuked for, then winking conspiratorially at me while the PM praised their acuity.

And yet, despite such experiences, I clung to the ideals of the meritocracy. I told myself they were isolated incidents, 'just one bad apple', and not indicative of the whole industry. After all, I had found a job whenever I'd needed one, I had been given opportunities, and I had earned the respect of the people whose opinions mattered to me. I believed myself to be walking proof of the meritocracy.

Now I see that I was downplaying ordinary, everyday bias. I was discounting the energy I wasted on things my male colleagues didn't even have to think about.

Like always having to reassure customers that I was indeed the technician.

Like having to work that much harder, and speak that bit louder, to prove I knew what I was talking about.

Like being criticised, sometimes very harshly, for my deficient 'soft skills', while no one cared when the men didn't have them either.

I could go on. What strikes me now is not the frequency of these events, but how I accepted them. I chose to ignore the background static of everyday bias because I *needed* IT to be a meritocracy. Before I reached working age, I had already endured a lifetime of punishment for things I couldn't control. I *had* to believe I'd found a place where I was judged solely on what I *could* control. On some deeply buried level, I equated meritocracy with safety.

*

Eventually, my first SyncID project kicked off. I was raring to go, to build a great solution and to make my mark—but there were problems from the start.

Firstly, the project had been very badly sold. SyncID was on the hook to deliver a solution at a fixed price while another company did the business analysis and produced the requirements. So basically, we had to build something yet to be defined and without any control over the extent of that definition. For a contractually fixed price.

The project manager had started with SyncID more recently than me and had never managed a commercial project of this type. I don't know what he spent his time doing, but it was definitely not controlling the scope. So I tried to do it, arguing every point with the customer in an effort to limit the blank cheque we had given him. Eight fractious and frustrating months later, the customer finally realised that a never-ending project didn't make him look any good and he reined in his demands and agreed to compromises. I was able to deliver something, and while I wouldn't say it was a solution I was exactly proud of, it functioned.

It had been a tough project, with a PM that had been more of a hindrance than a help, so it irked me to hear that, within SyncID executive, he was getting all the credit for the successful delivery. This was a pattern I was to see repeated again and again: project managers were over-credited for success, but the flipside was that they were also over-blamed for failure. I would come to believe that this skewed apportioning of merit was a big reason why the company had a revolving door on the PM team. Training yet another rookie PM in the realities of identity projects became an unacknowledged part of my job.

At SyncID I missed the well-considered management practices and deliberate fostering of a positive company culture that I'd enjoyed at LANExpert, and I found none of the project templates and methodologies I'd been trained on at AusCM. With little competition in the identity space, SyncID were getting plenty of work, but the company still had the feel of a startup with processes that were murky, immature, and made-up on the fly.

There were red flags, but I'd had such high hopes coming in that I didn't want to give up too soon. I decided to frame the issues I saw as an opportunity—I would make SyncID better! I would point out the problems when I saw them, I would develop the missing project resources myself, and I would back up the voices of colleagues who were likewise trying to improve things. I resolved to push, cajole, and lead by example until SyncID became the company that I wanted to work for.

*

At the same time, I was keeping up my MVP activities. A month after I moved back to Australia, I presented at TechEd 2011 in Australia and New Zealand, and as a full-time Identity specialist now, I had plenty of material for blog posts. On the 1st of January 2012, I was awarded my fourth consecutive MVP.

Later that year, I was again booked for TechEd, and this time, the organisers of the Australian event asked me to speak at the Women in Technology Lunch.

Seven years had passed since TechEd 2005 in Amsterdam, but the memories of the Lunch I'd attended there still rankled. I accepted this invitation, but first I confirmed that the attendees would have plenty of time to chat to each other before listening to me.

And so it was that I returned, as it were, to the scene of the crime. I was determined to right that earlier wrong by delivering the sort of talk I would have given anything to have heard back in 2005. I wanted to tell a positive story, one that might encourage younger women, especially any suffering pre-forty anxiety as I had back then, to stick with their career in tech.

With the Pollyannish title of 'My Brilliant Career, and why IT's a Great Career for Women!', my talk was essentially a highlight reel of my work history which now spanned twenty-seven years, counting back to that Apple job when I was sixteen. I skipped across the peaks: learning my trade, adventures in commercial IT, starting my family, international moves, dot-com silliness, flexible work, my first MVP award, and returning to Australia to my dream job as a MIIS consultant. I added some photos to enliven it, and I thought I'd done a pretty good job of demonstrating the 'great career' sub-title.

After my talk, the first audience question was one I probably should have seen coming, but in my resolutely optimistic mindset, was completely unprepared for.

'Can you tell us,' asked a woman, who looked to be around my age of early forties, 'about a time you experienced bias in IT, and what you did?'

'I haven't experienced bias in IT,' I stated.

There was an audible intake of breath, not just from that woman but from others around the room. I could see heads shaking in disbelief, so I doubled down.

'OK that's not quite true, the way I was hired at the dot-com because I was female was biased, but apart from that, I can't think of anything. The way I see it, if you have the technical skills that a customer or an employer needs, they don't care if you're a woman.'

The moment passed, I answered some less controversial questions, and the lunch ended.

Later, a young woman approached me asking if she could get my opinion on something. We went outside and sat in the warm afternoon sun, and she told me about her job as a programmer.

'I really love programming,' she said, 'but the guys at work are always putting me down. My boss gives me all the boring jobs, I don't think he believes I can handle anything more complicated.' Her voice was starting to crack, and I could see she was fighting back tears. 'I don't know if I can take it anymore, I think I'll have to find something different to do!'

I felt like an idiot. Here I'd been with my glib assertions about tech being a great career for women, and at least one person in the room was dealing with this bullshit. I felt awful for her, and angry on her behalf, but all I could think to say was, 'That sounds like a bad workplace, and I don't think there's much you can do about that. There are better ones out there, start looking.'

It wasn't a solution to the real problem, but I'd still give the same advice today. When you're the person being persistently undervalued at work, you're unlikely to be in any position to change the company culture.

Ironically, I didn't even recognise that in my goal to change the culture at SyncID, I had no more chance than she did.

*

You may have heard what Tolstoy said about unhappy families being unhappy in their own way? Well, you could say something similar about identity management projects: the projects I did with SyncID were each uniquely difficult in their own uniquely difficult way.

It wasn't because of the technology—that was the least of my worries. Between unrealistic expectations and bizarre machinations on the customers' side, combined with the pressures and inconsistent management on the SyncID side, almost every engagement became far more difficult than it needed to be.

I'm going to take you into one of them now, just to give you a little taste of how crazy things could get.

TUESDAY 9TH JULY 2013, CANBERRA

'What,' snarled the tech lead as soon as she saw me, 'have you DONE???'

It is a quarter to eight in the morning and I have just arrived. I'm exhausted after working fifty hours in the last four days, and I have no idea what she's on about.

'Our email, it's completely messed up!' she shouts. 'Why did you change everyone's email address??'

'MIIS doesn't even touch the email address…' I protest, but she's having none of it.

'Just fix it!' she snaps and marches off.

I sit down at my desk and login to my terminal, fretting over how long it's going to take—what with their convoluted login process—to get onto my server and figure out what's wrong. I can't imagine what might have happened. My MIIS solution only went live yesterday, and it performed perfectly all day. And like I told her, it doesn't even set the email address! With all the wrangling over requirements we've been through, she should know that!

Damn this slow login. How much time have I wasted in this place just waiting for logins to complete?

I swear that tech lead has it in for me. It's sad because I was so pleased to be working with a woman for once, but she's always had this reserve, and my buddy overtures were clearly not welcome. I took the hint and kept it all business, and for the first six months we worked together fine. But something changed a month ago. She became critical and demanding, and I have no idea why.

Right, I am finally logged in. Now to get onto the jump box.

I just don't get what her problem is. The solution was delivered on time and does everything they wanted. I even accommodated a last-minute requirement she flung at me. It was clearly out of scope and meant huge amounts of extra work over the weekend, but the PM said I had to do it.

That PM… Another woman I was excited to work with until things got real. She works for Microsoft because it's their project, and SyncID is the sub-contractor here. I'm the one doing all the work, but I'm answerable to this PM, and I can see now she's never been on my side.

Around the time the tech lead started turning on me, the project meetings got weird as well. Random customer people showed up wanting to know why the project wasn't on track, why it wasn't delivering 'value for money'. We'd reassure them that testing was going very well and we were right on schedule, but something was clearly up.

If the PM knew anything, she wouldn't tell me. Instead, after these meetings, she lectured, 'On a Microsoft project, we don't just aim to satisfy the requirements, we aim to *delight!*' Then she'd take off, leaving me none the wiser about how I'm supposed to *delight* customers who suddenly seem intent on finding fault with everything.

OK, I'm finally through all the hops and password gates and I'm on my MIIS server.

What's this? It's busy exporting a lot of changes to Active Directory, what's it doing? Oh, man, I can see what happened–this is so crazy!

I call over the stony-faced tech lead.

'Look at this! Someone in HR changed every single person's preferred first name to 'Cleanup'! The actual word 'Cleanup'! I can see they did it after close of business yesterday, and they reversed it this morning, so now everyone's proper names have come back in.'

She's still glaring wordlessly at me, so I explain further.

'MIIS doesn't care what the name is, it just takes whatever HR gives it and updates the user accounts. It changed them all last night… and, yep, it's just finished putting everyone's names back to what they should be. What a bizarre thing for HR to do! And on the first day we went live as well!'

I look beseechingly at her, hoping for some crack in her demeanour. I mean c'mon, everyone was briefly named CLEANUP! That's funny right? Give me *some-thing, woman*!

But no, she refuses to be drawn in, not ready to let me off the hook just yet.

'Why did *your* solution change all the email addresses?' she demands.

'Oh, well, that will be your email system,' I tell her. 'It must have updated the addresses after the names were changed, but don't worry, the old addresses are still there, and email will still be delivered. You just need to get someone to run the email address policy again and it will come good.'

She pauses a moment, like she's still trying to find a way for this all to be my fault. Then, without another word, she turns and stalks off.

'Oh, and then go talk to someone in HR and explain why they can't do things like that now!' I call towards her retreating back.

I spend a few minutes checking through logs, making sure everything else is running as expected. Apart from this Cleanup debacle, everything looks great. And even that was simply MIIS doing what it was supposed to do. This is data-driven automation, and the system is going to use the data it's given. We've had endless conversations about this, including with HR, so what did they think would happen?

I shake off my frustration and turn to check my email. The first one I see is from Ms Stony Face herself. She sent it just before I arrived and it's cc'd to everyone remotely connected to the project: within her organisation, at Microsoft, and a couple of higher-ups at SyncID. She says:

> I am writing to report that, after yesterday's go-live, the MIIS solution has caused a serious incident which is currently impacting our email and other critical corporate systems. The SyncID consultant, who has not yet arrived this morning, has clearly failed to adequately test and monitor the solution. We want to make absolutely clear that this is completely unacceptable.

Well fuck.

I try calling the PM, but she doesn't answer. Not that I expect she'll be any help—she'd probably just berate me for not being sufficiently *delightful*. I don't dare call anyone at SyncID because I fully expect them to blame me too.

I'm still sitting there wondering what on earth I should do when a follow-up email arrives from the tech lead. She starts by admitting that the issue was caused by someone in HR, but then concludes with a further attack on me:

> I accept that MIIS was just doing what it was programmed to do, however this is a common known problem that the SyncID consultant should have been aware of.

So let me get this straight: it's common for someone to change every single person's first name to 'Cleanup' in MIIS's primary data source, and I should have known that would happen? Sheesh.

I'm still digesting this latest character assassination when its author appears by my desk.

'You can go now,' she says.

'But I'm supposed to be here all week for post go-live support!' I protest.

'Just,' she growls, 'Get. OUT.'

*

When I reflect on experiences like that one, I wonder how I ever believed that IT was a meritocracy. Yes, I got jobs, but that's only part of it. If you're being harshly criticised over minor and even imagined faults, while at the same time your considerable efforts and successes go unacknowledged, you are hardly being treated according to your merit.

Over the years I worked for SyncID, I learnt something very annoying about merit: the better you perform, the less anyone notices. Your merit becomes assumed, just a mundane part of you that needs no comment, like your hair colour or your coffee order.

Perversely, if someone's performance swings between utter brilliance and near disaster, their moments of merit stand out to the point where they may be applauded, even celebrated. And while these people are feted in excess of their track record, the reliable performers are taken for granted.

It makes sense because it's how our brains work. Unless we're really paying attention, we notice what stands out and not what is there consistently. Without rigorous and unbiased measurement, we are terrible at judging merit.

Not all projects were like the Cleanup one. I enjoyed good relationships with many of my customers, but that allowed me to absolve the meritocracy. The Cleanup Project happened in an organisation known for its toxic workplace culture, and it was easier for me to write off that experience as an awful workplace than to seriously engage with the idea of a systemic problem. Like, for instance, gender bias.

The saddest thing about the Cleanup Project was how thrilled I'd been to work with both a female tech lead and PM, for the first time in all my years in IT. But when the going got tough, they both got mean. I wonder, would they have taken their frustrations out on a man in quite the same way? Would the tech lead have spoken so venomously and sent such defamatory emails? Would the PM have instructed a man to *delight*?

I don't blame them, well not much anyway. They were also operating within hierarchies that were so male dominated we all looked like novelties. They were dealing with it as best they could.

I came across that tech lead again a couple of years later. The Cleanup organisation had been merged into another, and I had been brought back in to modify my solution. She was there, looking more anxious than stony faced this time, as she was not at all happy about the merger.

Weirdly, she seemed to think we were *friends*. She came to me more than once looking for sympathy and asking my advice about searching for a new job. She never apologised about her behaviour on the Cleanup Project, and I didn't trust her anywhere near enough to bring it up.

Nor did I talk to her about what I had learnt since that project wrapped. Apparently, there had been a secret goal, formed at a higher level but never communicated to the project team. The MIIS solution, it was believed, would directly lead to two full-time positions being cut from the Service Desk.

Because the team was unaware of this, no analysis was ever done on workloads in the Service Desk and what might feasibly be automated. Had they done that, they would have quickly realised that MIIS doesn't replace people. Its job is to handle the type of fiddly and labour-intensive tasks that weren't being done anyway, because no one had the time for them. MIIS was only ever going to assist the Service Desk—it doesn't do their jobs.

Late in the project, some big boss man had come asking how soon after go-live he could start firing people. And he'd been told, I imagine with a great deal of surprise and consternation, that there was no chance of that happening. He was most displeased.

This explained the mood change. It explained the people who showed up unannounced, wanting to know why the project wasn't delivering 'value for money'. It was why the Microsoft PM insisted I go above and beyond, and why the tech lead blamed me as hard and fast as she could. I had copped it from everyone, not because I'd done anything wrong, but because the project's merit was suddenly being judged against entirely different criteria and they needed a fall guy.

IT never was and never will be a meritocracy—it is too human for that. We judge merit, not rationally against universally agreed criteria, but with emotion and bias and priorities that can change in a second.

I see now that I clung to the meritocracy myth for so long because it gave me hope. When people were dicks, I believed that the quality of my work would win them over in the end. But who was really benefitting when I worked myself beyond exhaustion, risking damage to my physical and mental health, all in the name of the meritocracy? Not me, that's for sure.

In 2014, Microsoft CEO Satya Nadella spoke at a major event for women in technology. He suggested that women should *not* negotiate their salary but instead place their trust in tech industry employers to make it right once they had proven their merit.

Nadella's speech occurred two years after my own 'there is no bias' gaffe at TechEd, but had I been in the room in 2014, I might still have bought Nadella's argument. Fortunately, there were plenty of women there who'd had quite enough of this meritocracy bullshit. There was an immediate and impassioned backlash on social media, prompting Nadella to apologise for and retract his statement.

As a brown-skinned man who had climbed to such heights, Nadella may, like I had, have seen himself as walking proof of the meritocracy. One data point, however, is not proof. It is an example, and it may be an outlier.

Tech is not a meritocracy, but that doesn't mean we should stop fighting for a world where merit outweighs bias. The data has long been in: overcoming bias doesn't just benefit those of us who look different, it benefits everyone.

26. How to find a new groove

CANBERRA IS A LOVELY PLACE TO LIVE, COMBINING a neighbourly suburban feel with the amenities and employment opportunities of a much larger city. Our kids quickly settled into their schools and made friends, and Nathaniel and I had short commutes and plenty of time for them and each other. A year in, after Nathaniel's first Canberra job ended, I tempted him over to the dark side and a new career in IT as a Business Analyst.

We had a good life in Canberra, but after three years, it still felt tenuous—like this was another posting and we would soon be moving on. So, we decided to dig in properly by terminating the lease on our rental and buying a house. We even, satisfying a long-term plea of Pascale's, adopted a cat. Canberra felt more and more like home.

Our lives settled into comfortable grooves revolving around school, work, family, and friends. I found the time to take up violin again, enlisting a teacher and, for the first time since the age of eleven, playing classical music. I even joined an orchestra. At first, I could only sit at the back of the second violins and air-bow half the time, but it was nonetheless exhilarating to be amidst the music as it swelled and whirled around me.

Canberra had given me some great new grooves, but at the same time I was still circling some tired old ones that went all the way back to childhood. Particularly at work.

My deeply embedded primary school routines repeated in the way I endured toxic environments, brushed off micro (and sometimes macro) aggressions, and pushed through by putting my head down and working. Some might call that resilience, but I now call it trauma conditioning. Whatever you call it, it's stress, and eventually I was going to have to reckon with the toll.

At the same time I was in another groove that was so habitual I couldn't even see it. At SyncID, my job was to be the reliable consultant who delivered my projects successfully and without complaint. Or, as I'd been trained so young, the good girl with the good grades who quietly dealt with her own needs. In this groove, I was grasping to meet a standard of 'bare minimum' that, no matter how hard I worked and how much I achieved, always remained just out of reach.

And that's hard work. In 2017, after six years and many projects with SyncID, I was well on my way to burning out.

*

There are two parts to burnout: what you have too much of, and what you don't have enough of. I had work stress that was excessive and prolonged, and I didn't feel reliably supported or appreciated.

By 2017, instead of the anticipation and enthusiasm with which I'd once approached each new project, I now felt something closer to dread. I still enjoyed the technical and problem-solving parts of my work, but the rest of it—the politics and the frustrations and the finger pointing—were killing me. I wasn't quite ready to give up on all the hopes I'd had for this company, and on all that I'd invested in trying to improve our delivery methods, but something had to shift.

Once upon a time, I had made a small change in my work practice, a simple tweaking of my availability to users, which had a hugely positive impact on my stress levels and job satisfaction. Perhaps I could do something like that again, something to break me out of this cycle.

I decided to cut my working hours. Not the ones I was paid for, but all the ones I wasn't.

I had long argued about overtime with SyncID's management team, but according to the CFO we were well paid enough for unlimited overtime to be part of the package. I had, on many occasions, made it clear I didn't want *money*, I just wanted *time*. I knew as well as anyone that out-of-hours work was part of IT, but I still deserved time off in recompense. The CFO wouldn't hear it. To me, his refusal of something as basic and cost-free as time-in-lieu communicated clearly that he cared nothing about my health and well-being.

Well, if he wasn't going to give me what I needed, I was going to take it. From then on, I worked my forty hours and then I shut my laptop, turned off my phone alerts, and went dark. And if I had to work weekends or evenings, I made it clear to the project manager that I would be taking equivalent time off to recover. What were they going to do, fire me?

They didn't, but in giving up unpaid work I fired myself—not from SyncID, but from my other job as an MVP. As my tolerance for overtime evaporated, so too did my enthusiasm for my MVP activities. For years I had been keeping up my blog, running a monthly online meet-up for MIIS practitioners, and helping out on open source projects. That had all been great... until suddenly it wasn't. In 2017 the internal engine that had been driving me all that time ran out of fuel. When I came up for renomination for the 2018 award, even though it would have meant a nicely round-numbered tenth MVP, I declined.

The year ended with its usual Christmas-New Year, summer shutdown. The break was good, but whenever I thought about returning to work in January, I felt heavy and tired. Cutting overtime hadn't been enough, I needed to try something else.

Perhaps, I thought, I could repeat what happened when I'd first discovered MIIS at London Business School. Perhaps learning something new would reignite my passion for tech. When I returned to work in January 2018, I rang the Professional Services Manager and told him what was going to happen.

'I'm enrolling in an online course on machine learning,' I said, 'and I am going to drop down to four days a week for the next six months to do it. I don't expect anything from SyncID, I will pay for the course myself, and those days can be considered leave without pay. Just to be clear, I will not be available on my study days. This is not up for discussion.'

SyncID had just as much of a revolving door with the Professional Services Manager as it had with the PMs and this guy hadn't been in the job for long, but he was a good sort, and he assured me he'd make it happen.

For the next six months I spent my Wednesdays learning about algorithm generation, supervised versus unsupervised learning, and matrix mathematics. It was fun, but it didn't make the four days of my working week any better. After the course, when I reverted to full-time, I felt just as flat as before.

I was working on yet another 'difficult' job, thanks to certain people in the customer organisation who seemed determined to derail the project. Nonetheless, I liked the team I worked with, and I'm too much of a completionist to leave a solution unfinished, so I battled through. When I got to the end I knew I was done. There was no way I could face another of those gruelling projects.

It was now the end of 2018, and I had been with SyncID for over seven years. In that time, I had worked on some fantastic projects with fantastic people doing work I was genuinely proud of. But too many had been in toxic environments where political machinations were allowed to trample effective change, and that was depressing and exhausting. Especially as I was representing a company I no longer believed in.

Replaying another of my patterns, I handed in my notice without first lining anything up. All I wanted was to take a long break. My notice period would end in early 2019, and Nathaniel and I were planning a trip to Japan for my 50th birthday in July. Perhaps, I thought, I'd spend the intervening months practising my violin, planting things in my garden, and hanging out with my teenaged kids. There'd be plenty of time to look for my next gig later in the year.

But news of my resignation got around fast, and before I'd finished with SyncID my phone was already ringing.

*

The first call was from Simon. I'd worked with him a couple of years earlier when I'd been the SyncID consultant and he'd been the customer's Enterprise Architect. I knew he'd gone on to found an IT services company with a group of other consultants, and that they'd named it A23 after the main road that brings you into Canberra.

'I've heard you're leaving SyncID,' said Simon, 'and we'd like to offer you a job.'

I was touched. I liked Simon a lot, and those of the other A23 founders that I'd met as well, but it felt too soon.

'I've promised myself a break right now,' I said. 'I really appreciate you thinking of me, but I've had years of representing other people's companies, and honestly, I'm not sure I want to do that again.'

The next phone call was from Farryn. She'd been the customer's project manager at the same place I'd met Simon. Only twenty-two at the time, she'd impressed me with her brains, her resourcefulness, and her knack of hassling people into doing what she needed in a way that left them completely charmed. Farryn had since moved to another Government Department, and that's what she was calling me about.

'We really need you here,' she said. 'They're starting an Identity project, but they already had a bad experience with a vendor, and they need someone they can trust. I've told the CIO about you, and he wants to meet you.'

It didn't feel right to knock Farryn back after she'd talked me up to this CIO, and I guessed it wouldn't hurt to just meet for a coffee...

Within ten minutes of sitting down with the CIO, he'd snagged my interest.

'There's a lot of problems with IT here, but I've assembled a great team and we're working hard at uplifting infrastructure and maturing our processes,' he told me. 'I don't know what went wrong with the old Identity Management system, they switched it off before I got here, but we really need something. The vendor is telling me that a simple software upgrade is all we need to reactivate it, but I'm not sure. If it was that easy, why hasn't it happened already? I need someone who really understands this sort of thing to look into it and tell me the truth.'

'You'll definitely get that from me,' I said.

It turned out that this role was the very one that Simon had been trying to recruit me for. I phoned him.

'OK,' I said, 'I'm in. But I have conditions. I want to be a contractor, not permanent. I want to set my own work hours and schedule. I will not work with anyone I don't want to, and I will have full freedom to speak my mind.'

'No problem, you do whatever makes you feel comfortable. We're just happy to have you,' he replied.

After so many years of accepting whatever I was offered, I was finally in a place where I could say what I needed.

*

I started contracting through A23 in early 2019, working as an Identity Management consultant in the Department where Farryn had introduced me. I worked for them full-time, analysing their needs and building a new solution that suited them far better than the unreliable one they'd switched off.

By the end of the year, they had a working Identity Management system which did far more than their old one had. At this point, had I still been working for SyncID, the project would have wrapped and I would have left, taking all my hard-won knowledge with me. This was, I realised, a groove I had never liked. When I enjoyed working with a customer, as I had with this one, I became invested. I wanted to stay engaged, to extend my solution, and to adapt it as their needs changed.

At the same time though, there was another groove I wanted to avoid: the one where I get taken for granted.

I spoke to my customer about dropping down to one day a week. 'It's all you really need now,' I assured them. 'The solution is built and it's just about maintaining it and extending as new requirements come up.'

My second argument was the clincher, and it came out of bad memories of another groove—the one I'd got stuck in at Money Bank.

'There's always delays when introducing something new,' I told them. 'So we'll only be able to move slowly on extending this solution anyway. If I just come in one day a week, we can move at a steady pace and you're not paying me to wait around for other people.'

It was the perfect compromise for us both: they'd retain my expertise, at a fraction of what they'd been paying, and I'd get to go and do other things.

I already had options for the other four days of the week. My MIIS skills were in short supply and high demand, and

I was constantly turning away offers. I spoke to two organisations who had recently been in touch and offered them each two days a week, still contracting through A23.

In early 2020, I began settling into my new groove, one in which I work across multiple long-term customers. I was getting to know my two new customers, remotely as they were both in other states, and I was still supporting my existing one in Canberra. This way of working was, I was confident, going to suit me just fine.

*

In February, I read worrying reports about a 'novel coronavirus'. Travel restrictions were rumoured when, in March, I flew to Western Australia to meet one of my new customers. After that, my planned trip to Tasmania to meet the other one was cancelled. Nathaniel and the kids were sent home, and the first of many lockdowns began.

We now had five people in the house since my mother-in-law had moved in with us a few years earlier, and with everyone home all day, it was feeling cramped. By sheer good luck, we had an almost-completed studio in the backyard that we'd built to be a spare bedroom, teenager hangout, and music room. It didn't have proper flooring yet, but it had power and a nice view of the garden. I took a desk, a chair, and my collection of laptops, and I moved in.

We settled into our Covid bubble and, between the masks, the handwashing, and the paranoia about every sniffle or scratchy throat, we made the best of it.

I would have been happy in my little Canberra groove, but I had someone else to worry about. My mother still lived alone in northern Sydney, and she was as up and down as ever. She was already isolated enough, and the pandemic was only going to make that worse. I phoned often, and visited as conditions allowed, but eventually, as I knew it would, her next breakdown came. This time, with no one around to manipulate, she had to call me herself. She sounded terrible.

As I'd done with her so many times in my life, I went straight into rescue mode, dropping everything and driving four hours to her place. I found her cowering in the dark, with food rotting in the fridge. She'd turned off the main electricity switch for the house so the scammers wouldn't get her through the power points.

I bundled her up and took her back to Canberra. Without going into the details, the end result of the weeks she spent with us was that she agreed to sell her house and move somewhere with better support.

'I'd like to live round the corner from you,' she said.

'Not on your life,' I thought, and guided her towards a beautiful retirement community in a historic and well-resourced country town, not far away.

I'm not going to go through everything it took to get a mentally unwell woman in her seventies out of her home of forty years during a global pandemic, I'll just say it was a *lot*. I hired in what help I could, but there was no one to whom I could outsource the daily emotional labour of keeping her on track through her mood swings, changes of mind, and random accusations against everyone involved.

At the start of 2021, she was settled in her new home, and I was a wreck. If I'd been burnt out towards the end of my time with SyncID, it was nothing compared to this. I was emotionally, psychologically, and physically drained. At the same time, my usual ways of refilling my cup, live music in particular, were unavailable.

I was already triggered enough when a particularly nasty and accusatory email from my father tipped me into my own mental health crisis, the one I mentioned way back at the start of this book. This was the crisis that would eventually propel me into a daily writing practice, in which I examined my past and wondered how I fit into the history of women in computing. It was what led to me writing this memoir.

But that came much later. First, I had many weeks of visceral flashbacks, stabbing physical pains, and washing machine brain in the most intense and unrelenting spin cycle I'd ever experienced. I knew I needed help, but so did half the country and psychologists' books were full. So, I helped myself with one of the most reliable grooves I've ever had: I learnt everything I could about my condition.

I read all about childhood trauma: *The Body Keeps the Score* by Bessel van der Kolk, *Adult Children of Emotionally Immature Parents* by Lindsay Gibson, *What My Bones Know* by Stephanie Foo, and many others besides. I learnt how trauma had affected the development of my brain and nervous system, and how this damage had been playing out over the course of my life.

To calm my nervous system, I went daily to YouTube for healing doses of yoga and NPR Tiny Desk Concerts. I subscribed to reputable mental health podcasts, and I listened to them on walks in a nearby reserve. I forced myself to be present for my family, to check in on them, to listen to them properly, even when half my brain was screaming in anguish.

But, as one of my podcasters often said, you can't heal your broken brain from inside your broken brain—you need someone outside it to help. Eventually, through word of mouth, I was lucky enough to find a therapist who specialises in adult recovery from developmental trauma. It was then that this latest phase of my healing began.

Part of that healing has been writing this book. In doing so, I've figured out a lot about what drove me, and how the puzzle pieces of my life fit together. It's been enormous amounts of work, and some of the memories were extremely hard to revisit, but many reminiscences of my career have been wonderful.

I chuckled as I remembered Geoff with his bare feet, and Lew's gruff fatherliness, and Laura telling the guys 'I don't buy it' if they tried to bamboozle her with jargon. There are many more wonderful and quirky people who have educated, inspired, and encouraged me along the way—I couldn't include everyone, but I remember them all with fondness and gratitude. I may have enjoyed working with tech, but it's been the people who have made this a great career.

*

It is 2025 as I'm writing these final words—where has all the time gone?

I still contract through A23, and I still work across multiple customers, but I arrange it so I rarely do a forty-hour week. As promised, A23 management leaves me to do my own thing, but at the same time, they treat me as a member of the family. I'm that independent and slightly peculiar Aunty Carol, always disappearing off doing goodness knows what, but welcomed back with affection when I decide to make an appearance. It's the best of both worlds: I have an employee's sense of belonging, with a contractor's freedom and flexibility.

A23 was a startup when I joined them, but a realistically grounded and sustainable one. I have taken great pleasure in watching them grow from a gang of eleven plucky IT contractors into a busy and thriving company with a fantastic culture and their sights set on expansion across Australia. In 2024, Simon stepped back from the CEO role and the founders elected Natalie, the one woman among them, to take his place. I could not be prouder to be working for an IT Services company that is headed, not just by a woman, but by an Australian First Nations woman. Natalie is an absolute powerhouse and is leading A23 into a great future.

That young firecracker of a PM, Farryn, is also blazing ahead. Eight years after I met her, she is now the Chief Information Officer of a government agency. I am ridiculously proud of her.

It finally feels like things are changing. As I walk around my customer's IT departments there are many more women than there used to be. They're still more likely to be doing non-technical jobs, but they shift the dynamic nonetheless.

I've also noticed a shift in the way people relate to me—they're more respectful of my opinions and not so quick to disagree. Maybe it's simply because I am now older than almost everyone I work with, however I think what I'm seeing is a broader acceptance of diversity in IT.

This year I started working with two new customers, and it still surprises me when they defer to my opinion, saying 'You're the expert!' Sometimes I catch myself arguing too hard to get my point across even though no one is disagreeing. 'I'm sorry,' I've had to say, 'I'm just so used to getting pushback on this.'

Microsoft announced that MIIS will be 'end of life' in 2029. That just happens to be the year I turn sixty. Once upon a time I could not imagine retiring, but now I think that maybe MIIS and I will retire together. Until then, I'll stay in this comfortable work groove that I designed for myself.

For so much of my career, I was the odd one out. I could never really be 'one of the guys', and I felt too different to most of the women to fit in with them either. But, as the Riot Grrrls showed the world, you don't have to fit in to make something great. Like they did, I have taken ownership of my damage, and harnessed my anger, and combined it with my passion and creativity to forge my own version of success. Considering all that was stacked against me, I feel like I've won.

Just before I sign off, there is one last thing that I promised you right back at the start of this book. That band I put together, the IT Grrrls, how did we go in the competition?

27. How to be an IT Grrrl Part 2

SUNDAY 10TH DECEMBER, 2023, CANBERRA

'My name's Carol…' I start, before the crowd cuts me off with cheers and applause. I cling to the microphone stand, looking down and laughing as I wait for them to stop.

I've already introduced the rest of the band.

Charmaine: 'Our Power BI expert and chanteuse extraordinaire!!'

Kiara: 'She's got the voice of an angel and the coding skills of a demon!!'

Monica: 'She can build a website and keep the beat!!'

Sharon: 'She controls the IT finance and the groove!!'

Serin: 'She manages the IT service delivery and all those sexy chords!!'

Together, we are the IT Grrrls, and we've made it to the final of the Department of Rock, Public Sector Battle of the Bands.

As in the earlier rounds, I introduce myself last. Usually I say, 'My name's Carol, and I'm an Identity and Access Management Solutions Architect, and if you don't know what that is, don't ask me because *I will tell you*'.

That always got a laugh. But tonight, I have something different I want to say.

The applause has died down, and everyone is looking at me expectantly. I can feel my emotions welling and I swallow them down, willing my voice to remain steady.

'My name's Carol,' I start again, 'and I am incredibly grateful and proud to be standing on this stage with these five amazing women. Just a few short months ago when I saw the post about this competition, I thought, "How cool would it be to put together a band made up entirely of women who work in IT?" I didn't know if it was even possible–it's not exactly a large pool to draw from. Somehow, we found each other, and these women have shown up with enthusiasm and talent and so much joy throughout this whole project. It has been an absolute blast!'

My Grrrls grin happily at each other, clapping along with the audience. I don't know if they have any idea just how huge this moment is for me.

It feels like all the parts of me have come together: the bullied and friendless Smellbomb, the punky early twenties Carol with her screwdriver and her indie rock band, the earnest but argumentative IT professional, the perpetual odd one out who sought solace and community in live music, the mother encouraging her children to be themselves, the geek who loved to talk about tech…

In this moment, on this stage, in front of this audience who has laughed and clapped along with my IT-skewed lyrics, I am proud to be all of those things.

'This is our last song of the competition,' I say into the mic, 'and we think you'll really like it.'

I place my 5-string electric violin under my chin and look towards Monica, who taps her sticks to count us in. I play the intro, a jolly, folky riff. Then Charmaine sings:

> My phone pings, just after three
> It's another recruiter, wants to know if I'm free
> I know I really am the fortunate one
> 'Cos I can make my code run, yeah I can make
> my code run

At our first band meeting, we'd joked about making it to the final, but even now, almost at the end of our set, I can't quite believe it really happened. At this point, it doesn't matter whether we win—we got here, and we did it together. And now all that is left is to get the whole audience singing along with us:

> Girls just wanna write code
> Girls just wanna write code
> Girls just wanna write code!

You know it.

Acknowledgements

I NEVER IMAGINED I WOULD ONE DAY WRITE AND publish my own memoir, and bringing it into existence would not have been possible without generous friends and talented professionals.

Thank you, Simon and Michelle, for the many reviews, comments, and words of encouragement you gave me when I was just getting started. Thank you Tania, Emma, Jemal, Wayne, Kathryn, Margaret, Jen, Charmaine, Diana, Mary Jane, Hannah, Shane and Jane for reading and providing feedback on drafts and excerpts.

Huge thanks to my writing coach and editor Liza for everything you've taught me about storytelling and for becoming so invested in mine. Thank you to Alicia for your perceptive cover design and to Bev for shepherding me through the self-publishing process.

And the biggest thanks of all to Nathaniel, my partner, greatest supporter and love of my life.

www.ingramcontent.com/pod-product-compliance
Lightning Source LLC
LaVergne TN
LVHW100253210926
848480LV00003B/912